DEVELOPMENTS IN MEDICINAL MUSHROOM
BIOLOGY AND THERAPEUTIC PROPERTIES

The Editors

Prof. (Dr.) K.K.Janardhanan, Ph.D, FNSE, FNABS, Graduated from University of Kerala, India. Obtained Master's and Ph.D degrees from University of Rajasthan Jaipur, India. He worked as a Post-Doctorate Fellow in the Department of Biogenesis, Institute of Microbiology, Prague, as Fellow of Czechoslovak Academy of Sciences.

He was Deputy Director, Central Institute of Medicinal & Aromatic Plants (CSIR, Government of India), Lucknow, before joining Amala Cancer Research Centre, Thrissur. Currently, he is Professor, Department of Microbiology, Amala Cancer Research Centre, Thrissur, India. He has been working on the medicinal mushrooms occurring in India for the last 15 years for the development of therapeutically useful products for the treatment of human diseases.

He carries with him over 40 years of R & D experience. His main areas of interest are medicinal plants, medicinal mushrooms and mycomedicine. He has published over 200 research papers in peer reviewed National and International Journals and 4 books. He is a Fellow of National Academy of Biological Sciences and Fellow of National Society of Ethnopharmacology. He was the President of Mycological Society of India and is currently the Vice-President of National Society of Ethnopharmacology. He is a recipient of Dr. Agnihothrudu Memorial Endowment Oration Award for his outstanding contributions on medicinal mushrooms. He was recently conferred the Bharath Jyoti Award by India International Friendship Society, New Delhi, for his Meritorious Services.

Prof. (Dr) T.A. Ajith, Ph.D, Received Master of Science in Medical Biochemistry with first rank from Manipal Academy of Higher Education, Manipal, Karnataka, India and Ph. D in Medical Biochemistry from University of Calicut, Kerala, India. He is a Professor in Biochemistry at Amala Institute of Medical Sciences, since July 2012. He has authored 57 original research papers in peer-reviewed National and International journals with 875 citations and 6 chapters in text books. His research is primarily focused on the role of compounds on the mitochondrial energy status. Dr. T. A. Ajith serves as Editorial Board member for 12 National and International journals and reviewer for many journals. He is an active Lab Assessor for NABH, Quality Council of India (QCI) and co-principal investigator of several research projects.

DEVELOPMENTS IN MEDICINAL MUSHROOM
BIOLOGY AND THERAPEUTIC PROPERTIES

——————— Editors ———————

K. K. Janardhanan, Ph.D, FNSE, FNABS
Professor
Department of Microbiology
Amala Cancer Research Centre
Thrissur - 680 555, Kerala, India

T. A. Ajith, Ph.D
Professor
Department of Biochemistry
Amla Institute of Medical Sciences
Thrissur - 680 555, Kerala, India

2015
Daya Publishing House©
A Division of
Astral International Pvt. Ltd.
New Delhi - 110 002

© 2015 EDITORS

ISBN 978-93-5130-659-7 (International Edition)

Published by   :   **Daya Publishing House®**
                          A Division of
                          Astral International Pvt. Ltd.
                          – ISO 9001:2008 Certified Company –
                          4760-61/23, Ansari Road, Darya Ganj
                          New Delhi-110 002
                          Ph. 011-43549197, 23278134
                          E-mail: info@astralint.com
                          Website: www.astralint.com

Laser Typesetting :   **Classic Computer Services, Delhi - 110 035**

Printed at   :   **Replika Press Pvt. Ltd.**

PRINTED IN INDIA

PREFACE

Mankind is on constant search for new bioactive substances which can be useful remedy for the ailments affecting the people. The Western society recently has given great emphasis on natural products as a source for the development of healthcare products. About 3.5 billion people worldwide, well over half of world population depend on plant based medicines for their primary healthcare. Majority of new drugs that have been generated in recent years were from natural products. Plants, fungi and bacteria have been the major sources for new drug discovery. Antibiotics, antimalarials, immunosuppressants and anticancer drugs such as penicillin, quinine, cyclosporine and taxol respectively were developed from plants and fungi.

Mushrooms are macrofungi. Currently, 14,000 to 15,000 species of mushrooms are known in the world and among them nearly 700 have been known to have significant pharmacological properties. Medicinal mushrooms have an established history of use in traditional medicine. The early herbalists were more interested in medicinal values of mushrooms than their use as a source of food. Contemporary research has validated and documented the ancient knowledge. Many bioactive substances with immunomodulating properties have been isolated recently from mushrooms. Medicinal mushrooms have been demonstrated to possess a number of therapeutic properties including antitumor, immunomodulation, antioxidant, radical scavenging, cardioprotective, hypocholestrolemic, antiviral, antiparasitic, antifungal, hepatoprotective, antidiabetic, and adaptogenic effects. However, medicinal mushrooms are vast and largely unexplored source of powerful new pharmaceutical products.

The current volume entitled 'Developments in medicinal mushroom biology and therapeutic properties' is a treatise aimed at providing insights to the biodiversity, biology, cultivation, food and nutritive value, ethnopharmacology, chemistry of bioactive molecules, medicinal properties, and therapeutic uses of medicinal mushrooms particularly occurring in India. All the chapters have been written by scientists engaged in productive research in the area of edible as well as medicinal mushrooms. We are sure that the book will be of immense use to students, teachers, scientists as well as to pharmaceutical industry. We hope that several chapters addressing the therapeutic potential of medicinal mushrooms will inspire modern clinical practitioners to use medicinal mushroom products for treatment. We thank authors for their outstanding contributions to assemble this treatise. We are also thankful to M/S Astral International (P) Ltd for the excellent printing and production of the book.

K.K. Janardhanan
T.A. Ajith

FOREWORD

The role and significance of natural products, herbal medicines are being increasingly appreciated in recent years for the prevention and cure of several human diseases. There is an increasing awareness of the significance of ethnic and traditional knowledge in the development of therapeutics. Medicinal mushrooms have an established history of use in traditional Oriental medicine. Medicinal effects have been demonstrated for many traditionally used mushrooms. Modern clinical practice in China, Korea, and other Asian countries continue rely on mushroom derived preparations. Traditional Chinese medicine has been practiced in a clinical manner although pharmacological bases are not well understood.

Tropical forests of India has a rich wealth of mushroom flora. Mycodiversity and fungal wealth of this ecosystem can be expected to be a 'store house' of new and novel pharmaceutical products. Extensive studies have been carried out in India on the cultivation and production of a number of medicinal mushrooms. This was focused on the possibility of popularizing the cultivation of medicinal mushrooms. However, investigations on the pharmacological properties and their therapeutic use have largely ignored. In this context I am indeed happy to know that Amala Cancer Research Centre, Thrissur, Kerala, India organized the first National Symposium on Medicinal Mushrooms in India. This symposium was able to project the prospects and utility of medicinal mushroom research in India.

Eminent scientists and professors who are associated with mushroom research have contributed to the proceedings. Their articles deal with various aspects of mushroom research such as ethnomycology, ethnopharmacology, biodiversity and conservation of medicinal mushrooms. Chemistry of medicinal some wild mushrooms is an added attraction of this volume. Another important aspects contributed to the volume are various pharmacological properties of a variety of mushrooms, product development, practical application and safety evaluation of selected commercial products. Cultivation methods of several medicinal mushrooms such as *Ganoderma lucidum*, *Lentinus edodes*, *Pleurotus* species have also been adequately looked in the proceedings.

I must congratulate the organizers of the symposium for bringing out such an interesting compilation on medicinal mushrooms entitled '**Developments in medicinal mushroom biology and therapeutic properties**'. I am sure this will be a useful publication not only to scientist perusing research on medicinal mushrooms in India but also to the international research community.

Professor Solomon P. Wasser
Institute of Evolution & Department of Evolutionary and Environmental Biology
University of Haifa, Haifa 31905, Israel
Editor-in-Chief
International Journal of Medicinal Mushrooms

MESSAGE

Mushrooms are edible, delicious and potentially medicinal. They contain a large number of biologically active chemical compounds in their fruiting bodies and cultured mycelia. Medicinal mushrooms have an established history in folklore and traditional medicine in several countries in the world. Contemporary research worldwide has validated and documented this ancient knowledge which led to the development of a number of products with unique properties from medicinal mushrooms in the last three decades. Modern clinical practice in Japan, China, Korea, Russia, Malaysia and several Southeast Asian countries extensively use mushroom-derived preparations.

Majority of new drugs in modern medicine have been generated from natural products. Mushrooms are largely unexplored source for drug discovery. The production of mushroom derived compounds is restricted because their availability is mostly confined to their fruiting bodies and mycelium. Hence, development of proper technology for the cultivation of medicinal mushrooms is of significant importance. In this context, the current publication entitled 'Developments in Medicinal Mushroom Biology and Therapeutic Properties' is a commendable contribution to the medicinal mushroom research and development. I am happy to see that a large number of experts engaged in the research and development activities of medicinal mushrooms have contributed to the volume. I hope this publication will be highly useful to students, teachers, scientists and medicinal mushroom industry as well. I congratulate the editors for their invaluable contribution.

P.L. Sunil Kumar
Director
Daehsan Trading
Dubai, U.A.E

ACKNOWLEDGEMENTS

We are indebted to Rev. Fr. Walter Thelappily CMI (Former Director) and Rev. Fr. Francis Kurissery CMI (Present Director), Amala Institute of Medical Sciences, Thrissur for their support and encouragements for organizing the first National Symposium on Medicinal Mushrooms in India at Amala Cancer Research Centre, Kerala, Thrissur. This volume is a compilation based on the presentations by a large number of participants in the symposium.

We are grateful to our colleagues, Dr. Ramadasan Kuttan, Dr. Jose Padikkala, Dr. Girija Kuttan, Dr. T.D.Babu, Dr. C.R.Achuthan of Amala Cancer Research Centre, for their support.

We are also thankful to all contributors for their valuable and informative contributions to this volume. Special thanks are due to Professor Solomon P. Wasser for contributing a foreword to this book.

We are indebted to DXN- Daehsan Trading (India) Pvt. Ltd, Chennai, for their generous financial support.

We sincerely express our gratitude to the type setting service by Mr. Jananrdhanan Kallayil, MM Creation service, Thrissur, Kerala and M/S Astral International (P) Ltd, New Delhi for the excellent printing and publication of the book.

K.K. Janardhanan
T.A. Ajith

CONTENTS

LIST OF CONTRIBUTORS

Abraham Salees P
6414 Misty Creek
Missouri City, Texas
USA

Ajith Kumar Bordoli
Principal Technical Officer
CSIR – North East Institute of Science and
Technology
Jorhat
Assam

Ajith. T.A
Professor
Department of Biochemistry
Amala Institute of Medical Sciences
Thrissur, Kerala

Anjum Varshney
HAIC Agro Research & Development Centre
Sonipat, Hariyana

Baruah Gitarthi
CSIR –North East Institute of Science and
Technology
Jorhat
Assam

Bhat Gulzar A
CSIR- Indian Institute of Integrative Medicine
Kashmir –Srinagar

Bhattacharyya. P.R
CSIR –North East Institute of Science and
Technology
Jorhat
Assam

Dato' Dr.Lim Siow Jin,
Chairman & CEO, DXN
Malaysia

Deshmukh Sunil Kumar
Senior Group Leader
Department of Natural Products
Mumbai

Dhar B.L
Mushroom Research Development and
Training Centre
DK Floriculture, New Delhi

George Varughese
Director, Amity Institute of Phytochemistry and
Phytomedicine
Thiruvananthapuram, Kerala

Himanshu
Mushroom Research Development and
Training Centre
DK Floriculture, New Delhi

Ijinu. T.P
Amity Institute of Phytochemistry and
Phytomedicine
Thiruvananthapuram

Janardhanan. K.K.
Professor
Amala Cancer Research Centre
Thrissur, Kerala

Jitindra
Mushroom Research Development and
Training Centre
DK Floriculture, New Delhi

Joseph Soniamol
Tropical Botanic Garden and Research Institute
Palode, Thiruvanantapuram, Kerala

Kaviyarasan. V
Professor
Centre for Advanced Studies in Botany
University of Madras
Chennai, Tamil Nadu

Krishnakumar. N.M
Tropical Botanic Garden and Research Institute
Thruvananthapuram Kerala

Kumar Jitendra
HAIC Agro Research & Development Centre
Sonipat, Hariyana

Latha. P.G.
Director
Tropical Botanic Garden and Research
Institute
Thiruvananthapuram, Kerala

Lulu Das
Professor
Department of Plant Pathology
College of Agriculture, Vellayani
Thiruvanathapuram, Kerala

Makkar Sujata
HAIC Agro Research & Development Centre
Sonipat, Hariyana

Mathew. A.V
Professor
Kerala Agricultural University
Regional Agriculture Research Station
Kumarakom
Kottayam, Kerala

Meera. C.R
Assistant Professor
Department of Microbiology
St. Mary's College
Thrissur, Kerala

Nitha. B
Assistant Professor
Sree Ayyappa College
Alappuzha, Kerala

Pandey Meera
Principal Scientist
Indian Institute of Horticultural Research
Banglore, Karnataka

Prathibha. P.R
Department of Plant Pathology
College of Agriculture, Vellayani
Thiruvanathapuram, Kerala

Priyenka
Mushroom Research Development and
Training Centre
DK Floriculture, New Delhi

Pushpangadan .P
Director General
Amity Institute for Herbal & Biotech
Product Development
Thiruvananthapuram, Kerala

Rajasekharan. S
Scientist in Director's Grade
Tropical Botanic Garden and Research
Institute
Thruvananthapuram, Kerala

Shawal Abdul S
Scientist Emeritus
CSIR- Indian Institute of Integrative
Medicine
Kashmir –Srinagar

Shwet Kamal
Directorate of Mushroom Research
Chambaghat
Solan, Himachal Pradesh

Singh Ajay
Principal Scientist
HAIC Agro Research & Development
Centre
Sonipat, Hariyana

Singh Manjit
Director
Directorate of Mushroom Research
Chambaghat
Solan
Himachal Pradesh

Smina. T.P
CeNTAB, SASTRA University,
Thanjavur, Tamilnadu

Sonika
Mushroom Research Development and
Training Centre
DK Floriculture, New Delhi

Srivastava Neeraj
Mushroom Research Development and
Training Centre
DK Floriculture, New Delhi

Sudheesh. N.P
Department of Science and Technology
New Delhi

Suja. S.R
Tropical Botanic Garden and Research Institute
Thruvananthapuram, Kerala

Upadhyay. R.C
Principal Scientist
Directorate of Mushroom Research
Chambaghat
Solan
Himachal Pradesh

Veena. S.S
Indian Institute of Horticultural Research
Banglore, Karnataka

Veena Ravindran K.
Department of Microbiology
Amala Cancer Research Centre
Thrissur, Kerala

Verekar Shilpa A
Scientist
Department of Natural Products
Nicholas Piramal Research Centre
Mumbai

1

My years of experience in practical application of medicinal mushrooms

Dato' Dr. Lim Siow Jin
Chairman and CEO, DXN Malaysia

Today, we have the first National Symposium on Medicinal Mushrooms in India. I am happy to participate in the symposium and to deliver the keynote address. This is indeed a great honour to me. When we talk about the medicinal mushrooms, have we ever wonder who are the first user of Medicinal Mushrooms? Some people say it is Japanese and some say it is Chinese. It is not. Indeed, it is the Indian Monks who first apply it. The Indian Monks while traveling to China preserved the mushroom as a source of food and medicine. They taught the monks in China and the China's monastery slowly taught all the people about the medicinal mushrooms. This is true in case of many ancient arts. When we see the Kung Fu fighting, people asked, who the originator of Kung Fu is. It is again Indian Monk. When Indian Monks went to China, they have to defend themselves and protect the Sutra. So they practiced the martial art. Later they taught the Chinese to become Kung Fu masters. When we use beautiful Bonsai, people asked who discovered the art of Bonsai. It is not the Japanese or Chinese, again, it the Indian Monk. They preserved the medicinal plant in bonsai form and brought it to China.

After traveling for thousand years outside India finally medicinal mushrooms came back to India. How effective are the medicinal mushrooms? I have been dealing with medicinal mushrooms more than 20 years and have seen many miracles and excellent application of mushrooms. Actually all mushrooms are medicinal. The oyster mushroom, shiitake mushroom, maitake mushroom, St. George's mushroom, and all mushrooms are medicinal in nature. If we are mushroom growers and if we want to use mushroom as a therapeutic substance, then one thing is very important. We should not only know the mushroom but also we should study the substrate, because the efficacy of the mushrooms depends very much on the medium that we grow on. For medicinal mushroom normally we use a richer medium.

Now coming to the cancer, many people are asking me, can *Ganoderma* cure cancer? A cancer patient will be suffering from pain, they will be suffering a lot, after chemotherapy; hair will fall, sometime they will have throat ulcer and will have so many problematic side effects. For cancer the first thing we are very sure is that, mushrooms improve the quality of life. We have many cases, after taking mushrooms with chemotherapy the hair fall is less or do not fall at all, they do not have throat ulcer, they do not have the negative symptoms and with this we can improve the quality of the life of a cancer patient. Of course, there is a lot of successful treatment for cancer patients with mushrooms.

The life quality of a cancer patient is actually important. We had one case in Malaysia. There was an old man having cancer near the ear and neck with a lot of smelly, watery discharge from the neck. The family members were changing the napkin in the neck every half an hour. So I asked them to apply the mushroom powder and after sometime the wounds dried up. When I met the patient, he held me "Dr. Lim, I am an old man and whether my cancer is cured or not, it is not important to me, but what makes me feel so good is that the family members do not need to change the napkin for every half an hour. My heart is at peace now". So cancer patients get lot of improvement in life quality after using mushroom. Secondly, how to treat the mushroom to have better effects? Many people are only interested in the composition of the mushroom and how to extract it. There was a professor from Singapore, the Singapore Government granted him to extract the polysaccharide from mushroom for cancer patients. After he got everything successful and patent, he came to see me. He said "Dr. Lim let us join hand to work on polysaccharides." I told him no, it will not work. It is costly to extract and the result is poorer than the un-extracted whole mushroom. Later he failed in the business as it is too costly for his products compete. People want to identify what is important in mushroom and how it cures diseases. If you want to do this commercially, I advice you, please do not go in that direction. Mushrooms are rich in micronutrients. Any single mushroom even the oyster mushroom, shiitake mushroom, if tested in laboratory we get easily over 300 types of chemical components. So why just extract one component and hope it will work?

After consuming *Ganoderma* may patients could lengthen their lives .The scientists are scratching their head asking how mushrooms can do all these things. We are also thinking this for the past 20 years. Our only answer is the micronutrients. Actually a healthy human being can produce many micronutrients needed for the body. But for certain people due to some reasons, they may not be able to produce certain types of micronutrients and hence they should get from the source of food. What are the foods that are rich in micronutrients, one is mushroom, another one is algae like *Spirullina,* and third is from fermentation of microbes. A simple fermentation of vinegar will produce 100 to 200 types of enzymes and nutrients. For now, how to educate the people to promote the usage of these nutrients? First from our side, we are trying to apply the whole mushroom. In DXN, we are not only applying the whole fruit body, we even cultivate the mycelium. By combining the two, we got fantastic results. Throughout the whole world untill today, we have more than 5 million people associated with us and in India now it is more than 1 million.

Why mushrooms are so miraculous? and why mushrooms are so good to have such an amazing effects? My study is concentrated on two aspects. One is detoxification and other is balancing the body for regeneration. The scientists say once the nerve is broken it never regenerate and once kidney is shrunken, it never comes back again. We prove it to be wrong. The doctors said many stroke patients had no chance for them to walk again, but after taking *Ganoderma* mushroom many are actively walking again. We have many cases where the kidney had already shrunken, urination is not possible, and the patients lost all their vitality, sitting in the wheel chairs, after taking *Ganoderma* they gained back the health .There was a bank manger in Indonesia his kidney is failed and he was very sick . As he was an important person he came to the bank, punched his card, brought all the works to his house, did everything and in the evening came back to punch the card again . He took an X-ray of his kidney, both the kidneys had already shrunken and doctors told him nothing can be done. After consuming *Ganoderma* for some time, they took photos again and the kidney started to grow back. He was supposed to go for three dialysis in a week and after consuming *Ganoderma* for two months, it reduced to two dialysis and after talking for one year it reduced to one dialysis per week and after two years he became normal and stopped the dialysis. So regeneration power of the mushroom is fantastic and it is not only happening in India, it is not only happening in Malaysia, it is happening everywhere throughout the whole world, in the Middle East, in Africa, in Europe, in America, etc. So the detoxification and regeneration can be achieved by the administration of *Ganoderma*.

The direct proof of detoxification and regeneration are available over the past 20 years; for example diabetic patients have gangrene on the legs and the doctor advice for amputation. After administration of *Ganoderma* and putting *Ganoderma* powder for detoxification on the gangrene, smelly liquid ooze out and it is a good sign. This does not mean that the wound had become more serious, initially the wound is blackish in colour and slowly it turns reddish, slowly it turns whitish and then returns back to the normal. It is the power of *Ganoderma,* the process of detoxification and regeneration is very important and we are able to observe it directly.

Many people have liver problem like cirrhosis and hardening of the liver, the liver returned to normal after the patient taken *Ganoderma* . In cardiovascular diseases, many people are supposed to go for bypass and before bypass mega dosage of *Ganoderma* is advised . Mega dosage means 30 grams of any dry mushroom, it can be oyster, it can be button, it can be *Ganoderma,* it can be *Cordyceps* . We have tried to so many mushrooms, they have almost the same effects. So those of you, who do not have *Ganoderma* in your place, do not need to worry. Oyster as well is doing good. Some people can consume 30 grams in one shot, some people will take it in two times, or some people will talk 6 to 7 times. It does not matter, as long as one consumes 30 grams in a day.

For treatment of serious diseases we go for mega dosage. Eventually we can see all patients undergo detoxification and they will get a lot of reactions, we have to be very careful. One day there was a telephone call from Singapore, it was an angry call. He said "every time I took your *Ganoderma*, I had stomach pain . What have you put into your *Ganoderma*? May be you have put some chemical or something that is making my stomach painful. What is wrong with your *Ganoderma* ? I answered him, "There are millions of people taking *Ganoderma* and they do not complain of the stomach problem. Our *Ganoderma* are organically cultivated and processed in GMP certified factory. There is nothing wrong with *Ganoderma* but something wrong with your stomach! You better go and check up your stomach. "Soon enough he went to the hospital for check and found that he had first stage of ulcer. Doctor told him he is luckly to check it up as most ulcers initially do not cause much problem, when they do, it is already serious . So when we study the medicinal effect of *Ganoderma*, we need to study the other reaction of the body. From the reaction, we know what is wrong with the body.

Coming back to cancer, for cancer patients it is important first to reduce the suffering of the patient. When you reduce the suffering at least they can live a normal life . Before we talk about the cure and treatment, there is one important experience to share with you .You see, any living thing when the body is beyond repair, they may die by their own. When a cat catches a mouse and the mouse knows that it is beyond escaping it dies. The same may work in human being. That is why those are treating the cancer patients with *Ganoderma* must know one thing .When the cancer patient is in a very weak stage or in the last stage, after we administer *Ganoderma*, many of them improve and then passed away peacefully, as contrast with patient without taking *Ganoderma*, suffer tremendously before they die . This should be considering as the welfare of the terminal cancer patient. Of course many of them got treated and became totally well.

Through our experience mushroom can go along with any discipline of medicine. So, the allopathy doctors go with us, ayurvedic doctors, homeopathic doctors, also the traditional Chinese medical doctors all can work together with us. Many doctors like mushrooms because mushroom is a complementary to other therapeutic system. This is a short introduction of mushrooms. Maybe today it not be possible for us to talk about all the good things of the mushrooms, but still we feel today that it is a good beginning. So, may be in the future we can meet regularly and we can have more knowledge about the mushroom. This will boost the confidence of mycology in India.

Actually in my experience, this symposium is not a small one, it is a big symposium with so many quality papers that we are having here . Actually before I came to this symposium, even though I am promoting mushrooms in India for many years,due to the lack of communication with the scientists, my impression was that maybe the mushroom science is not common in India and this symposium has changed my mind totally. I can say that the arts and science of mushroom is already well established in India. Now we see from the North to the South, from Cordyceps, from *Ganoderma* and other types of edible mushrooms, we have so many studies on medicinal mushrooms and so many methods of cultivation discussed in the meeting. So, I feel this is a very successful symposium and we should mark this day as a great day for the medicinal mushroom advancement in India. The next step is how to bring this technology to the masses, to the cultivators, to the markets . Many of you had discussed with me that it is difficult to sell the mushroom in India. Let me tell my experience in Indonesia. In Indonesia, when I started to promote oyster mushroom, in one day we can only sell 5 kg and not more than that because people do not know to eat and cook this mushroom . Today, after a few years of promoting, now the demand is 7 tons a day and believe me, it is not enough to the market. So the market of mushroom is very difficult to start, but once you start you can never stop it because people simply love mushroom after trying it. Once they love it they don't let go. I still remember about 15 years ago when we started to promote this *Ganoderma* in India people were telling me "Dr. Lim, you must be crazy because Indian people may not know anything about this mushroom." But today we have more than 1 million people taking this mushroom. The turnover of mushroom of DXN in India alone is about 150 crores. I believe after today's symposium, all of us should go back and try to commercialize the mushroom in your hand.

From the start of this symposium, we say that all mushrooms are medicinal .When you are back, no need to concentrate on *Ganoderma*, concentrate on whatever mushroom you get in your area, you try to cultivate it, try to promote it, then try to market it . I am very sure that the market will come up.

Today we are very happy that not only this symposium is remarkably successful and also we got great support from Amala Cancer Research Center. In the future we will try to promote more members to cultivate the mushrooms. All the people can work together so we can cut down the cost of the mushroom and I will assure professor, it would not be 1000 rupee for 30 grams. I think we can cut down the price substantially..

There is no need to worry about the market of mushrooms. DXN in Kerala, India alone as stated in the newspaper recently making 13crores /year and the whole India of 150 to 200crores /year. There is a huge scope in mushroom business. So this is something very important for us and mushroom is also a very good way to encourage the farmers to eradicate poverty.

DXN farm size is only 20 hectors but we produce a few hundred million dollar worth of *Ganoderma* mushroom for the whole world and I believe the same experience may happen in India. So, even the farmers with a small land size will be benefited from the cultivation of mushroom. We as a scientist should try to promote this art to the farmers. We should try to simplify the cultivation technology and transfer the knowledge to farmers.

So in short I must congratulate organizers of the symposium, it is a very successful one, believe me and it is not a small symposium . It has all the quality papers and it is a big symposium. I believe after many years, people will remember this day when it kick start the medicinal mushroom industry in India.

2

Mushrooms as food, dietary supplements and medicine-present status and future needs

Manjit Singh, RC Upadhyay and Shwet Kamal

Directorate of Mushroom Research, Chambaghat, Solan (HP),
E mail: directordmr@gmail.com

Inadequate quality food supply, diminishing quality of health and increasing environmental deterioration are the three key problems of our times. Mushrooms address these issues as these are high quality food, have medicinal values/ neutraceutical properties and help in bioconversion/ bioremediation. Depending on the type of mushrooms, mushroom can serve as food, act as supplement, or serve as medicine. Furthermore, mushrooms are commonly considered tools for converting waste into wealth as these are cultivated on agrowastes. In this review, we describe the present status and future needs of mushrooms as food, dietary supplements and medicine.

Introduction

In the last few centuries the human population has increased at an exponential rate. The human population was estimated to be about 0.5 billion in 1500 AD. It took 300 years for the doubling. It took only about 125 years for it to double again from 1 to 2 billion. It became 4 billion in another 50 years (i.e. by 1975). The current population of over 6 billion is likely to reach over 9 billion by 2050. Considering the growth of human population on our planet having limited resources, it may be difficult to meet all the human needs in near future and population growth may turn out to be a bigger threat than the climate change. In fact many factors causing climate change are direct effect of increasing population.

We are already passing through an age of human health crisis like AIDS (in Swaziland, 41% pregnant women infected; 31% children orphan), Cancer (5.7 lakhs deaths annually in US), Hypertension, Cardiovascular diseases, Diabetes, Obesity (2/3rd adults in US are obese), Viral disorders (mad cow disease in Europe, foot-and-mouth disease, Bird flu, SARS), etc. The 3 key problems of our times are, inadequate quality food supply, diminishing quality of health and increasing environmental deterioration. Mushrooms address these issues as these are high quality food, have medicinal values/neutriceutical properties and help in bioconversion/ bioremediation. Mushrooms are Fungi, mainly belonging to class Basidiomycetes and Ascomycetes.

Fungi are unique organisms. Nothing in biology makes sense except in the light of evolution said Theodosius Dobzhansky. Earlier view was that fungi originated from plants. However, fungi have unique traits, common with plants as well as animals. In a paper in Nature 443, 818-822 (19 October 2006) 'Reconstructing the early evolution of Fungi using a six-gene phylogeny', the group of 70 authors (from 35 institutes of 6 countries) analysed 200 contemporary species to reconstruct the earliest days of fungi and concluded that plants split from animals about 1.547 billion years ago; and fungi split from animals about 1.538 billion years ago, i.e. after a gap of about 9 million years. The other key evidences in support of fungi-animalia clade have been arrived at by analysis of protein sequences, biosynthetic pathways, cytochrome systems, mitochondrial genetic material, biochemical and structural features, glycoproteins, mode of nutrition, storage of nutritive material, cell wall structure, etc.

Mushrooms are commonly considered tools for converting waste into wealth as these are cultivated on agrowastes. The total agricultural wastes in our country are over 1500 million tonnes of which over 600 MT are crop residues (Fig.1).

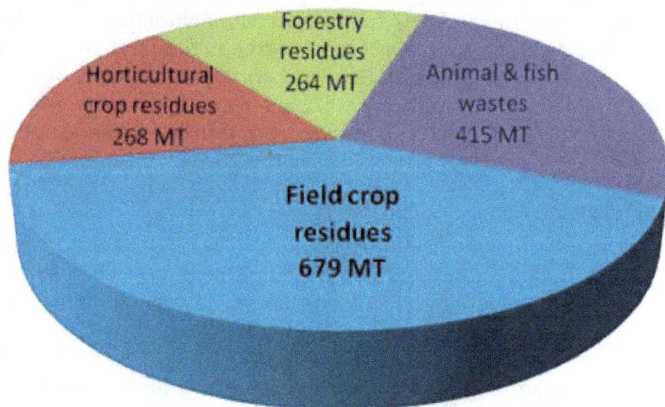

Total agricultural wastes in the country are 1566 MT
(NAAS, 2010)

Fig. 1: Estimated agrowaste production in different sectors in India

The second most important requirement for success in mushroom cultivation is manpower. The demographic pattern is going to change across the globe with rapid urbanization. However, in India despite rapid urbanization, majority in our country will continue to be in villages. In addition to this, India in 2020 will be young nation with average age of 29 years as compared to 48 years of Japan (Fig.2).

Despite rapid urbanisation, majority population in our country will continue to be in villages

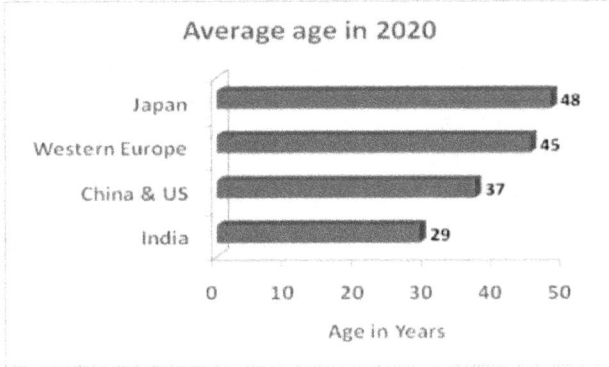

Average age in 2020

Country	Age
Japan	48
Western Europe	45
China & US	37
India	29

Age in Years

Fig. 2: Average age in 2020 in India and other countries

Producing mushrooms by using agrowastes not only address the issue of production of quality food and environmental issues (like pollution due to burning of straws, agri-residues), but can also address the health aspect as mushrooms are neutriceuticals and many of these have medicinal values. In fact depending on the type of mushrooms, mushroom can serve as food, act as supplement, or serve as medicine.

Since, inception of Directorate of Mushroom Research (earlier known as NCMRT/NRCM) in 1983, we have been collecting germplasm and at present there are over 2600 collections in our gene bank. Some of the medicinally important mushroom species collected and conserved in gene bank of DMR, Solan are given in Table-1.

Table 1: Medicinally important accessions in DMR Gene Bank

Genus	Species	No.	Genus	Species	No.
Grifola	frondosa	1	Isaria	synclarii*	1
	sulphureus	1	Schizophyllum	commune	11
Cordyceps	sinensis	3	Sparassis	crispa	5
Tremetes	versicolor	8	Auricularia	tenuis	5
	gibbosa	2		polytricha	6
	hirsuta	1		olivaceus**	1
	zonata	1		rosea*	1
	cinnabarinus	1	Hericium	coralloides	1
	pubescens	1		erinaceus	1
Phellinus	lloyadii	1	Pleurotus	tuber regium	1
	sublintenus	1		Others (16)	287

Contd.

Table 1 *Contd.*

	merilli	1	Termitomyces	robustus	1
	pectinatus	1		radicatus	1
	adamnetinus	1		mammiformis	1
	badius	1		heimii	1
Ganoderma	lucidum	42			
	tsugae	2	TOTAL		393

* First record from India ** New record for the world

Efforts have been made to cultivate these collections and some of the wild collections cultivated include *Auricularia* sp., *Hericium, Ganoderma* and *Pleurotus* spp. (Fig. 3).

Auricularia polytricha   Auricularia olivaceus

Fig. 3: Cultivation of *Auricularia* accessions available in DMR Gene Bank

The estimates of number of fungi expected to be present on the globe vary. From earlier estimates of 1.5 million fungi, today many believe that actual number may be more than 5 million. Only a small fraction of these have been identified. Mushrooms, for obvious reasons were among the first described fungi. In many cultures these were not only used as royal food, but their medicinal values were also well recognised. Efforts to cultivate these started few centuries back and there are reports on cultivation of *Auricularia* (600 AD), *Flammulina* (800-900 AD), *Lentinus* (1000-1100 AD) and *Agaricus* (1650 AD). However, scientific cultivation of mushrooms started in last century only. White button mushroom dominated in terms of contribution towards total mushroom production in early part of twentieth century. As per FAO stat the world mushroom production is over 6 m tonnes (Fig. 4).

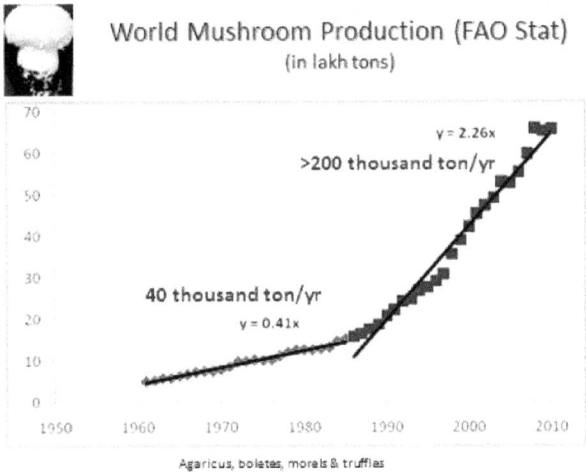

World Mushroom Production (FAO Stat)
(in lakh tons)

$y = 2.26x$

>200 thousand ton/yr

40 thousand ton/yr

$y = 0.41x$

Agaricus, boletes, morels & truffles

Fig. 4: World mushroom production as per FAO Stat

As per FAO stat code 449 the term mushroom includes *Agaricus, Boletus,* Morels and *Tubers*. Over time, number of other mushrooms have been cultivated and commercialized (Fig.5). China is reported to cultivate about 60 different type of mushrooms and considering all mushrooms, China alone is reported to produce around 22 m tonnes (Table 2). Considering that China is producing 80% of world's mushroom production, the global production can be expected over 25 MT. The growth in mushroom production, both of button as well as others has been exponential (Fig.4 and Fig.6).

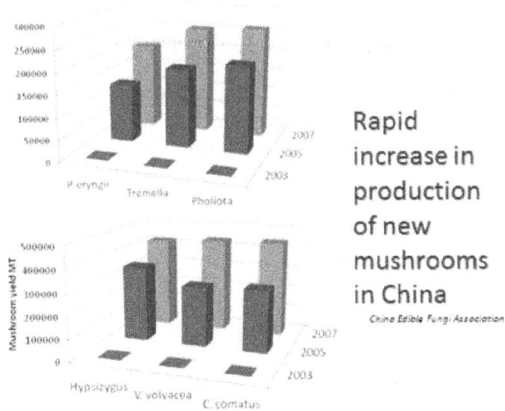

Rapid increase in production of new mushrooms in China

China Edible Fungi Association

Fig. 5: Increase in production of different mushroom in China since 2003

Table 2: Mushroom Production in China and world (all mushrooms)

Year	World ,000 ton	China ,000 ton	%
1978	1,060	60	5.7
1983	1453	175	12.0
1990	3,763	1,083	28.8
1994	4904	2640	53.8
1997	6,158	3,918	63.6
2002	12,250	8,650	70.6
2006		14,400	
2008		18,200	
2010		21,500	80.0 ?

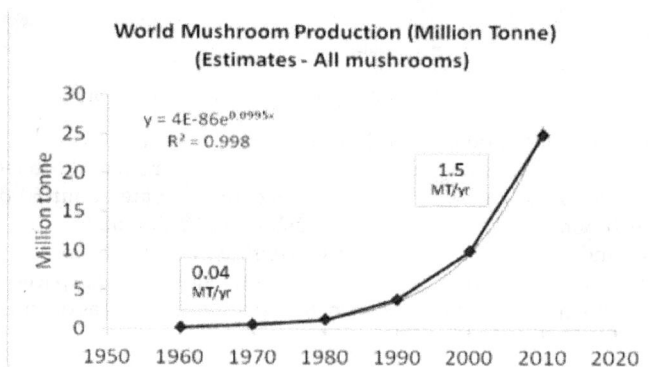

Fig. 6: Estimated world production of all mushrooms (million tones)

There has been a rapid growth both in production and diversification (Fig.5). From initial emphasis on production of mushrooms for food, there is a shift towards cultivation of mushrooms having neutriceutical and medicinal values. Button mushroom, that dominated the world mushroom production is struggling to be at number one position. Even in other mushroom species there are rapid changes in their production status. For example, *Lentinus*, that used to be number one mushroom in Japan, is now at third place and mushrooms introduced much later have taken over (Fig.7).

In India, mushroom scenario in the past five decades has undergone a sea change. With a modest beginning in 1960s, when India produced only few thousand tonnes of button mushroom in the hilly regions, today mushroom production has crossed one lakh tonne (Fig.8), there has been diversification in the mushroom species being cultivated and today mushrooms are cultivated in most parts of our country. Today, in addition to button mushroom, oyster, paddy straw and milky mushroom are being cultivated in tropical regions of our country (Fig.9).

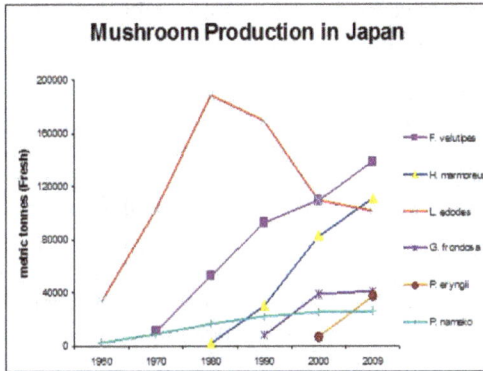

Fig. 7: Production of different mushrooms in Japan in last five decades

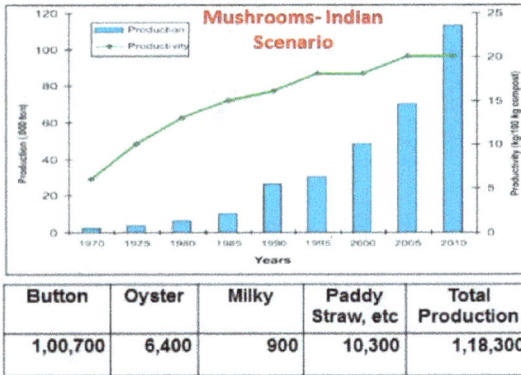

Button	Oyster	Milky	Paddy Straw, etc	Total Production
1,00,700	6,400	900	10,300	1,18,300

Fig. 8: Mushroom production scenario in India

Fig. 9: Mushrooms under cultivation in India (L to R *Agaricus bisporus, Calocybe indica, Pleurotus* sp. and *Volvariella volvacea*)

A few growers have also shown interest in cultivation of shiitake and other medicinal mushrooms. Presently, button mushroom is cultivated under both seasonal and controlled conditions and states like Haryana contribute more than 8,000 tonnes of mushroom by seasonal cultivation only. Part of button mushroom is also canned and exported. However, the export market has seen many ups and downs in the recent past. The other mushrooms are mainly grown under natural conditions and most of the produce is consumed locally. Some entrepreneurs have come out with innovative mushroom products that can help in growth of the industry.

Considering the availability of vast amounts of agro wastes, adequate labour, shift towards hi-tech agriculture and need for employment generation, particularly for youth, mushroom cultivation has a vast potential. The mushroom consumption in our country is about 0.04 kg as compared to 2-3 kg in Europe and more than 10 kg per person per year in China. There is a shift towards consumption of fresh mushroom in last few decades (Fig.10). This reflects the present status as well as scope for mushroom cultivation in our country.

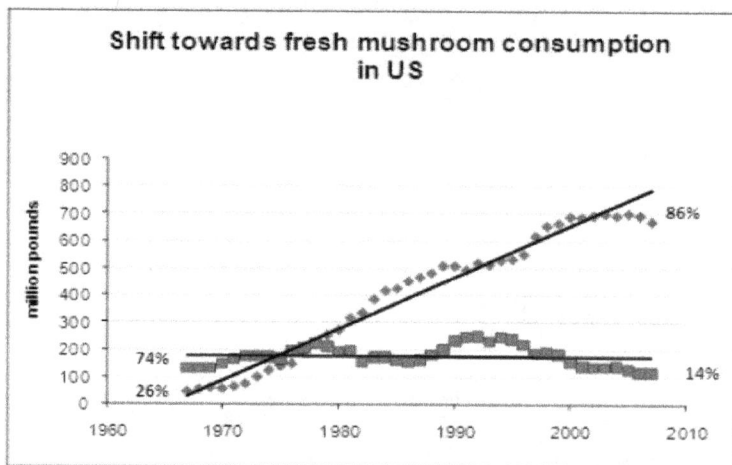

Fig. 10: Production of fresh and canned mushrooms in USA in last five decades

The growth in mushroom industry, however, is a function of positive interaction among researchers, extension workers, farmers, industry and policy makers. Hence, synchronized approach in systematically generating the awareness about mushroom consumption, proper technologies and environment for mushroom cultivation and development of marketing chains for supply of fresh mushrooms as well as production of indigenous mushroom products will be important.

Mushrooms as Food

Mushrooms are a source of quality protein having high degradability and all essential amino acids. Mushrooms are rich in fibre and vitamins, and are possibly the only vegetable source of vitamin D important to bone and muscles. Mushrooms are low calorie food and also rich in antioxidants. It is well established now that mushrooms are the health food. These are good for heart as mushrooms have little fat with highly polyunsaturated fatty acids (Fig.11), absence of cholesterol, and minimal sodium with rich potassium (Fig.12). Many mushrooms have anti-cholesterol compounds, such as Eritadenine (*A.blazei*, Reishi, shiitake), Lovastatin (Oyster), etc.

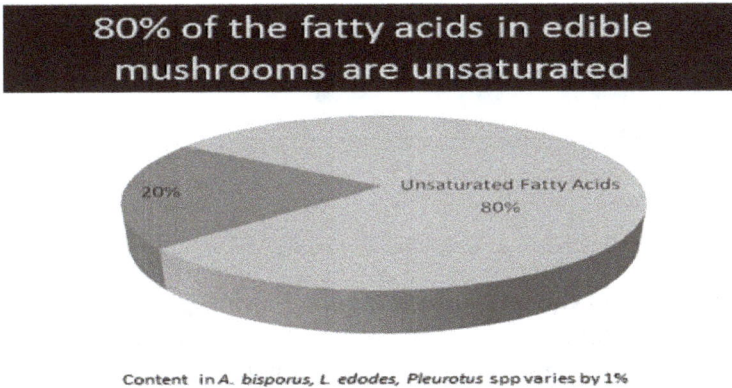

80% of the fatty acids in edible mushrooms are unsaturated

20%

Unsaturated Fatty Acids 80%

Content in *A. bisporus, L. edodes, Pleurotus* spp varies by 1%

Fig. 11: Proportion of unsaturated and saturated fatty acids in mushrooms

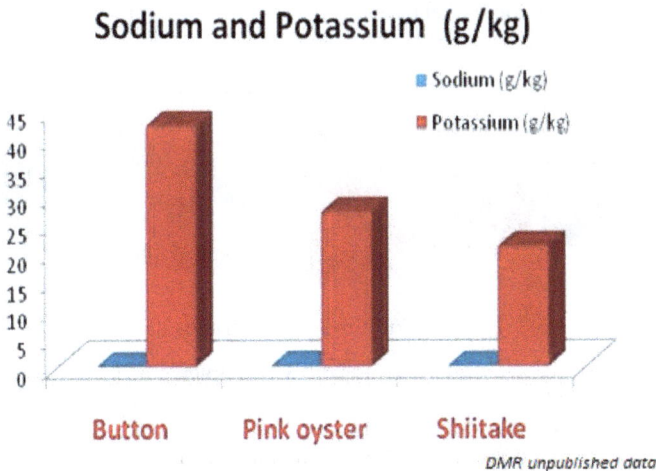

Sodium and Potassium (g/kg)

■ Sodium (g/kg)

■ Potassium (g/kg)

| | Button | Pink oyster | Shiitake |

DMR unpublished data

Fig. 12: Relative proportion of sodium and potassium (g/kg) in different mushrooms (DMR unpublished data)

Recent research proves mushrooms as a alternative to meat can help to manage weight. In addition, mushrooms also have adequate minerals and some of these are source of minerals like selenium considered to have anti-cancer properties (Fig. 13).

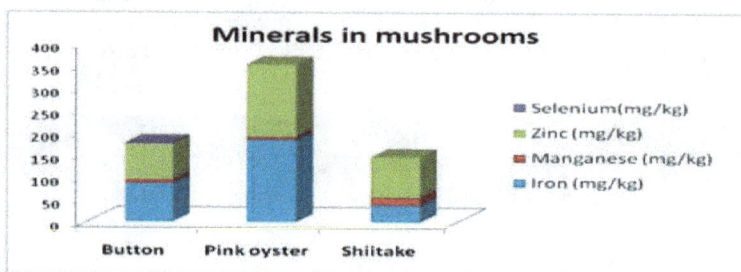

DMR unpublished data

Fig. 13: Minerals in different mushrooms (DMR unpublished data)

Mushrooms as dietary supplements

Dietary supplements have been described as ingredients obtained from foods, plants and mushrooms that are taken, without further modification, separately from foods for their presumed health-enhancing benefits. Mushroom derived preparations have variety of names like dietary supplements, tonics, functional foods, nutraceuticals, phytochemicals, food supplements, nutritional supplements, mycochemicals, biochemopreventives, designer foods, etc.[1] There has been rapid increase in trade of medicinal mushrooms and mushroom production that in 2006 contributed about 31% of world mushroom trade (Fig.14).

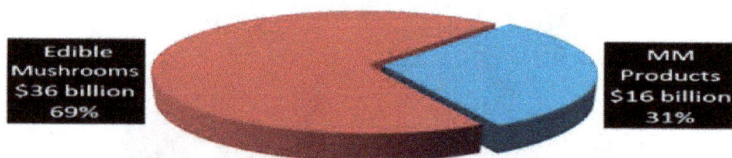

Fig. 14: World mushroom trade (2006)

There are various regulations for dietary supplements in different countries. For regulation of Health Foods in China, the National People's Congress passed the first comprehensive Food Safety Law (FSL) on 28th February 2009 wherein health food is defined as food processed for health functions. The rules granted consumers a right of action to sue for compensatory and punitive damages. In Taiwan Health Food Control Act of 1999, amended 2006, governs matters relating to health food products. Health food is legally defined as "food with specific nutrient or health maintenance effects, which are especially labeled or advertised, and do not aim at treating human diseases". The Mushroom Dietary Supplement Regulatory Framework in US requires facility registration, ingredients must conform to statute, product forms limited to oral use and are considered adulterated if a significant or unreasonable risk of illness or injury is present.

The standards vary in different countries and so do the claims of safety and quality. Any good product requires a uniform quality from batch to batch that is possible only when a uniform material is produced (which will require specific strains grown under controlled conditions) and is processed by following standard practices. The Newton's third law that for every action there is equal and opposite reaction applies every where – also when we tend to claim too many benefits of medicinal mushrooms. This leads to criticism, especially in the light of variable quality of products labelled identically. Thus, there is need for quality standards to have better market.

Mushrooms as Medicine

Many mushrooms that are used as food also have medicinal value. For example, oyster mushroom is a source of the drug class of statins (Lovastatin) used for lowering cholesterol and so preventing cardiovascular disease. Similarly, shiitake has Lentinan that is considered to have anti tumor, anti thrombosis, anti asthma, anti virus and anti cholesterol activity. Some of the mushrooms are still collected for medicinal use. The important example is *Cordyceps sinensis* that has Cordycepin (3'-deoxyadenosine, a derivative of the nucleoside adenosine) considered to provide energy and endurance.

Since, the discovery of penicillin in 1929 and streptomycin in 1943, there is desperate need to find new antibiotics – from new sources, as increased use has lead to spread of antibiotic resistance. Pleuromutilins are a class of antimicrobial compounds that inhibit bacterial protein synthesis and are active against a variety of pathogenic bacteria. Several different fungi have been reported to produce these compounds.

One area that has drawn global attention towards mushroom is their role in cancer and reducing side effects of chemotherapy and radiotherapy. Mushroom polysaccharides like β-D-Glucans linked to proteins have been tested on humans, as these are considered to enhance immunity. Immunoceuticals isolated from 30 mushrooms species have demonstrated anti tumor activity in animal treatments.[1] Some of the mushroom species, compounds and their likely effects are:

- *Lentinus edodes*- Lentinan, *T-cell oriented immuno-potentiators*
- *Grifola frondosa* – Grifon-D, breast, prostate, lung, liver cancer
- *Schizophyllum commune* –Schizophyllan, *T-cell oriented immunopotentiators*
- *Ganoderma lucidum* –GLPS polysaccharide fraction
- *Trametes versicolor* – PSK (Krestin), PSP (Polysaccharide peptide), immunostimulators
- *Inonotus obliqus* – Befungin
- *Flammulina velutipes* - Proflamin

- *Phellinus linteus* – Ethyl acetate fraction
- *Cordyceps sinensis* – Cordycepin, energy and endurance

Low-weight-molecular compounds considered to have medicinal value include Lectins, Terpenoids, Alkaloids, Antibiotics, etc. Medicinal mushrooms also contain enzymes like Laccase, Superoxide dismutase, Glucose oxidase and Peroxidases. Enzyme therapy prevents oxidative stress and inhibits cell growth.

Today cancer drugs are the biggest category of drugs in terms of sales worldwide. About 860 cancer drugs are being tested on humans. About 20% of Pfizer's, more than $ 7 billion R & D budget is on cancer research and 22 of about 100 drugs being tested by this company are anti cancer drugs.[2]

Modern medical practice relies on highly purified pharmaceutical compounds whose activity and toxicity show clear structure-function relationships. Herbal medicines contain mixtures of natural compounds, detailed chemical analyses are not always available and mechanism of action is not fully known. How to translate traditional Eastern practices into acceptable evidence-based Western therapies is an important step to promote mushroom as medicine. There is a long list of claims of medicinal benefits which need to be substantiated scientifically. A number of mushrooms are still collected from the wild and thus variations in chemical constituents are bound to be there. For any mushroom to be used commercially for medical use it is important to identify the strains with higher output of desired chemical, to standardize the cultivation and environmental conditions for maximizing the production of such constituents, to ensure adequate availability of raw materials and to have testing and evaluation procedures in place for proper labeling and traceability.[3] There is also need for generating greater information on stimulation of immunological systems and exact mode of action. Till then the status of medicinal mushrooms will be same as of numerous medicinal plants available in our country. Drugs, particularly those related to cancer treatment, are required in quantities that cannot be met by mushrooms quantity available. Many of these compounds are polysaccharides and their chemical synthesis is not easy. Many of the mushrooms are parasitic and their cultivation can lead to other problems. Commercialization will require growing fungal mycelia or fruit bodies under controlled conditions where we get uniform high quality product in large quantities.

References

1. Chang S.T. and Wasser, S. P., The role of culinary-medicinal mushrooms on human welfare with a pyramid model for human health. *Int. J. Med. Mushr.,* 2012,14, 95-134.

2. Petrova, R.D., New scientific approaches to cancer treatment: can medicinal mushrooms defeat the curse of the century? *Int J Med Mushr,* 2012,14, 1-20.

3. James, T.Y., Kauff, F., Schoch, C.L., Matheny, P.B., Hofstetter, V. and Cox, C.J., Reconstructing the early evolution of fungi using a six-gene phylogeny. *Nature,* 2006,443, 818-22.

3

Reishi – *Ganoderma lucidum*: King of herbs, a medical wonder

K.K. Janardhanan

Department of Microbiology, Amala Cancer Research Centre, Amala Nagar, Thrissur – 680 555, Kerala, India. E-mail: kkjanardhanan@yahoo.com

In ancient Oriental traditions, the importance of several mushroom species has been stressed. Among them, Lingzhi or Reishi (*Ganoderma lucidum* (W.Curt: Fr.) P. Karst) is considered as the most important that had contributed immensely for the treatment of ancient population in South East Asian countries. Lingzhi was used to treat several disease conditions such as to improve intellectual capacity and memory, to promote agility, to lengthen life span, to relieve hepatopathy, nephritis, hyperlipidemia, arthritis, asthma, gastric ulcer, arteriosclerosis, leucopenia, diabetes and anorexia. A large number of bioactive chemical compounds could be extracted from the fruiting body, mycelium and spores of Reishi. The mushroom contains a wide variety of bioactive molecules which included triterpenoids, polysaccharides, nucleotides, sterols, steroids, fatty acids, proteins/peptides, and several trace elements. Polysaccharides and triterpenes are the two major groups of bioactive compounds present. Based on consolidated scientific evidence available so far, Reishi has significant health benefits and would soon become a new therapeutic agent for treatment of a wide range of diseases.

Introduction

The role and significance of natural products, herbal medicines and tribal medicines are increasingly appreciated recently for the prevention and cure of human diseases. There is an increasing awareness of the significance of ethnic and traditional knowledge in the development of therapeutics. A rich heritage and long tradition of using herbal medicines for healthcare exists in several countries including India.[1] Mushrooms have been considered of medicinal value from ancient days. Ancient Oriental traditions have mentioned the importance of several mushrooms. The use of mushrooms as food and medicine has been mentioned in one of the oldest medical treatise, *Charak Samhitha,* as early as 3000 B C. Medicinal mushrooms have history of use in traditional ancient therapies and contemporary research has validated the ancient knowledge. Recent reports show that a total of 126 medicinal functions, including antitumor, cardioprotective, hepatoptotective, and antidiabetic are produced by medicinal mushrooms and fungi[2].

Mushrooms are abundant in nature and they have worldwide distribution. The number of mushrooms currently available in the world is estimate at 14,000 to 15,000. In ancient Oriental traditions the importance of several mushroom species has been stressed. Among them, Lingzhi or Reishi [*Ganoderma lucidum* (W.Curt: Fr.) P. Karst] is considered as the most important that had contributed immensely for the treatment of ancient population in South east countries including China (Fig. 1). For over 4000 years, red mushroom *G. lucidum* has been recognized by Traditional Chinese Medical practitioners as the highest ranked of all herbs found in the Chinese Pharmacopoeia. The Chinese name of *Ganoderma* is Lingzhi, means Spiritual Potency'. The Japanese name for *Ganoderma* is Reishi. *Ganoderma* is regarded as King of Herbs. Dr. Shi-Jean Lee, the famous medical doctor of Ming Dynasty, strongly endorsed the effectiveness of *Ganoderma* against a variety of diseases in his famous book, Ban Chao Gang Moo (Great Pharmacopoeia). According to him long term consumption of Reishi/Lingzhi will build a strong body and assure a long life.

Fig. 1: Fruiting body of Reishi – *Ganoderma lucidum*

Reishi-*G. lucidum* has played an important role in Traditional Chinese Medicine solely or in combination with other medicinal herbs. In ancient folk medicine Lingzhi was used to treat several disease conditions such as to improve intellectual capacity and memory, to promote agility, to lengthen life span, to relieve hepatopathy, nephritis, hyperlipedmia, arthritis, asthma, gastric ulcer, arteriosclerosis, leucopenia, diabetes and anorexia[3]. For long, *Ganoderma* has been considered as a panacea in China.

Bioactive components

A large number of bioactive chemical compounds could be extracted from the fruiting body, mycelium and spores of Reishi- *G. lucidum*. Over 300 reports have been published on the chemical constituents of *G. lucidum* and related species. Approximately 400 bioactive compounds have been reported from the fruiting body, mycelium, and spores of *G. lucidum* [2]. The bioactive molecules include polysaccharides, triterpenes, ganoderic acids, phenols, aminoacids, lignin, vitamins, nucleosides, nucleotides, sterols, steroids, proteins, unsaturated fatty acids [4], inorganic ions such as Mg, Ca, Zn, Mn, Fe, Cu and Ge[5] However, polysaccharides and triterpenes are the two major groups of bioactive compounds present in *G. lucidum*.

Polysaccharides

Polysaccharides of Reishi are the most important components and represent structurally – diverse biological macromolecules with wide range of biological and physiochemical properties [6]. The importance of polysaccharides and protein-bound polysaccharides in pharmaceutical industry has received considerable attention in recent years. More than 200 polysaccharides including protein bound polysaccharides have been reported from *G. lucidum*. The major bioactive polysaccharides isolated from *Ganoderma* species are β 1-3 and β 1-6 D glucans (Fig. 2). The basic structure is β-1-3 D glucopyronan with 1 to 15 units of β-1-6 monoglucosyl side chain[7]. *Ganoderma* also contain heteropolysaccharides, glycoproteins, and a group of polysaccharides known as ganoderan A, B, C. There are qualitative and quantitative differences in bioactive polysaccharides obtained from fruiting bodies and products of liquid fermentation.

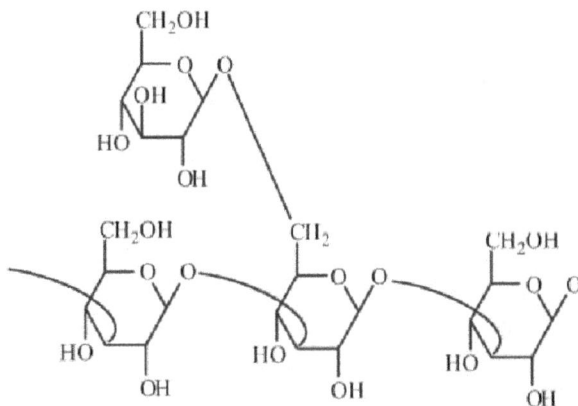

Fig. 2: Basic structure of *Ganoderma lucidum* polysaccharide

Triterpenes / Triterpenoids

One of the most important biologically active compounds in Reishi is triterpenoids/ triterpenes. This group has received considerable attention because of their well-known pharmacological activities. More than 130 oxygenated triterpenes (mostly lanostane-types) have been isolated till recently from the fruiting bodies, spores, mycelia, and culture media of Reishi- *G. lucidum*. These triterpenes could be divided into C30, C27, and C24 compounds according to the number of carbon atoms and based on the structure and functional groups (Fig. 3). In general, they have molecular weight ranging from 400 – 600 kDa. These compounds isolated from Reishi were intensively studied during 1980s. Currently, 136 triterpenes structure have been isolated from *G. lucidum* [2].

Other bioactive compounds

Reishi also contains sterols, amino acids, soluble proteins, oleic acid and ergosterol. A number of active proteins have been isolated from Reishi, while some proteins exist in the polysaccharide-protein complex. A novel protein, later named as Lingzhi-8 (LZ-8) was isolated from the mycelia extract of *G. lucidum*. Its complete amino acid sequence, biochemical and immunological properties have been described. In many plant and animal materials, bioactive peptides have been found to have antioxidant activity. However, *Ganoderma* peptides have been shown to be the major antioxidant component. Oxidative stress appears to be one of the major factors in the pathogenesis of many human diseases including cancer, aging and atherosclerosis. Various amino acids which are necessary for human body and high nutritional value and medicinal properties have been reported from Reishi. However, contents of aspartic acid, glutamic acid, alanine, and lucine from mycelia, spores and fruiting bodies were higher than other amino acids.

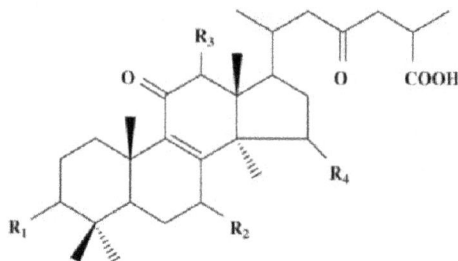

Ganoderic acid C_2 (1): $R_1=R_2=\beta$-OH, R_3=H, $R_4=\alpha$-OH

Ganoderic acid B (2): $R_1=R_2=\beta$-OH, R_3=H, R_4=O

Ganoderic acid AM_1 (3): $R_1=\beta$-OH, $R_2=R_4$=O, R_3=H

Ganoderic acid K (4): $R_1=R_2=\beta$-OH, $R_3=\beta$-OAC, R_4=O

Ganoderic acid H (5): $R_1=\beta$-OH, $R_2=R_4$=O, $R_3=\beta$-OAC

Fig. 3: Basic structure of *Ganoderma lucidum* triterpenes

Interest in cholesterol lowering properties of mushrooms has generated significant interest. Since then, these types of compounds have been isolated from ethanol extract of *G. lucidum* and the compounds were characterized by spectral analysis. The isolation and identification of 26-oxygenated sterols, ganoderol A, ganoderol B, ganoderal A and ganoderic acid Y and their effect on cholesterol synthesis have been reported. Applanoxidic acid, an antimicrobial compound was also reported to be isolated from Reishi.

Cyclo octasulfur and oleic acid were isolated from culture broth of *G. lucidum*. Both of these compounds inhibit histamine release which is an important activity for the treatment of allergies. Isolation of adenine, adenosine, uracil and uridine from the mycelium of *Ganoderma* species has been reported. Alkaloids, choline and betaine also were isolated from the spores of *G. lucidum*. The spores of the mushroom contain choline, betaine, tetracosanoic acid, stearic acid, palmitic acid, erosta-7, ergosterol and β-sitosterol[5].

Medicinal properties

G. lucidum has been used as a medical remedy in Traditional Chinese Medicine (TCM) and in many Asian countries during the past two millennia. This mushroom was thought to preserve human vitality and to promote longevity. *G. lucidum* was used to treat various human diseases such as allergy, arthritis, bronchitis, gastric ulcer, hyperglycemia, hypertension, chronic hepatitis, hepatopathy, insomnia, nephritis, neurasthenia, scleroderma, inflammation, and cancer. In addition to these *G. lucidum* has been reported to have a variety of pharmacological effects, which include immunomodulation, analgesic, anti-aging, antiatherosclerotic, antibacterial, antifibrotic, antinociceptive, and radioprotective (Table 1).

Table 1: Reported Medicinal properties of Reishi - *Ganoderma lucidum*

Fractions	Main Pharmacological Activities
Alcohol / Aqueous Extract	Anti oxidant Anti inflammatory Anti tumor Hypotensive Inhibition of platelet aggregation Cytotoxic to hepatoma cells Inhibition of histamine release Hypoglycemic Hepatoprotective Inhibition of cholesterol Inhibition of ACE Anti – HIV activity Anti - aging

Antioxidant activity

Antioxidant system is the body's defence mechanism against free radicals. Antioxidants are substances that protect vital molecules from the damages being caused by reactive oxygen species (ROS). Free radicals are chemical species capable of independent existence that contain one or more unpaired electrons. Important free radicals generated in the body are derivatives of oxygen and are termed as ROS. Examples are super oxide radical, hydroxyl radical, peroxy radical, hydrogen peroxide, singlet oxygen, nitric oxide radical, peroxynitrite, hypochlorous acid etc. Free radicals are involved in several disease conditions and in a number of pathological events. These radicals react with cell membrane and induce lipid peroxidation, cause extensive damage to DNA leading to mutation and carcinogenesis. Oxidative stress due to excessive production of ROS that out strip antioxidant defensive mechanism is a decisive etiological factor involved in a number of chronic human diseases such as cancer, cardiovascular diseases, inflammation, arthritis, diabetes mellitus, neurodegenerative diseases, and also aging. The protection of biomolecules from ROS mediated damage by various natural compounds is believed to be promising and practical approach to prevent cancer. Accumulating evidence from in vitro and in vivo studies has indicated that active components from *G. lucidum* have potent antioxidant and radical scavenging effects [8]. Experimental studies carried out in our laboratory have demonstrated profound antioxidant activity of ethyl acetate and alcoholic extracts of *G. lucidum*[9,10]. Polysaccharides isolated from *G. lucidum* also showed significant antioxidant activity. Recently two low molecular weight polysaccharides isolated from the mushroom also showed potent antioxidant activity (Table 2).

Table 2: *In vitro* antioxidant activity of *Ganoderma lucidum* extracts (IC$_{50}$ µg/ml)

Activities	Ethyl acetate extract	Methanol extract	Aqueous extract	Standard
Superoxide radical scavenging	213 ± 2.12	61 ± 2.5	475 ± 2.5	3.7 ± 0.16 (Quercetin)
Hydroxyl radical scavenging	185 ± 25	159 ± 3.6	140 ± 2	850 ± 20
Lipid peroxide inhibiting	205 ± 25	615 ± 4.08	-	418 ± 28.6 (Catechin)
*ORAC value	3894/100 g fresh weight			
All values are represented as mean ± S.D n = 3 * Micromoles of Trolox equivalent				

Anti-inflammatory and anti- arthritic activity

Inflammation, a fundamental protective response, can be harmful in conditions such as life threatening hypersensitive reactions to insect bites, drugs, toxins and in chronic diseases such as rheumatoid arthritis, atherosclerosis, lung fibrosis and cancer. Inflammation can accelerate the development of cancer and chronic inflammation is a risk factor for epithelial carcinogenesis. Prostaglandins generated during inflammation appear to be important in the pathogenesis of cancer.

Extracts of *G. lucidum* were demonstrated to possess significant anti inflammatory activity. The extracts were found to inhibit acute and chronic edema [11]. Several studies have confirmed anti-inflammatory activity of *G. lucidum*. The extract of the mushroom was also found to have profound anti nociceptive activity. *G. lucidum* has been shown to possess promising anti-arthritic activity. Boh et al[12] provided clinical evidence that demonstrated the effectiveness and safety of *G. lucidum* for joint health. In another study, Reishi compared favorably to prednisolone but with out the side effects. All these studies suggested that *G. lucium* has remarkable anti inflammatory, anti nociceptive and anti arthritic activities.

Anti-tumor and anticancer activity

Cancer is the second largest cause of death of children and adults, claiming more than six million lives each year worldwide. Tumor diseases are the main cause of cancer. *Ganoderma* species and their extracts are known as traditional remedy in Chinese medicine for the prevention and treatment of cancer[13]. Numerous pharmacological investigations demonstrated that hot water extract of *G. lucidum* inhibited tumor growth in several tumor-bearing mice. Several studies carried in our laboratory have demonstrated the significant antitumor effect of *G. lucidum*.[9,14,15] (Table 3). The alcoholic extract has been demonstrated to possess significant effect on mammary adenocarcinoma and skin papilloma [16].

Table 3: Effect of *Ganoderma lucidum* extracts on solid tumour

Treatments	Dosage (mg/kg)	Tumour volume (cm³)	Tumour weight (g)	% Decrease in tumour volume	% Decrease in tumour weight
Positive Control	-	1.053 ± 0.09	5.3 ± 0.76	-	-
Standard (Cisplatin)	4	0.046 ± 0.001	0.381 ± 0.060*	95	93
Aqueous extract	125	0.313 ± 0.010*	1.705 ± 0.090*	70	67
	250	0.182 ± 0.020*	1.248 ± 0.051*	80	76
	500	0.160 ± 0.010*	0.925 ± 0.004*	84.8	82.6
Methanol extract	125	0.375 ± 0.060*	1.960 ± 0.060 *	64	73
	500	0.167 ± 0.010*	0.921 ± 0.007*	84	82.6
Ethyl acetate extract	500	0.273± 0.030*	1.432 ± 0.028*	74.1	73
	1000	0.166 ± 0.034*	0.930 ± 0.035*	84.3	82.5
Values are mean ± S D ; n= 6 ; *P < 0.001					

The most important constituents of *G. lucidum* are terpenoids and polysaccharides. Both these constituents show remarkable antitumor activity [17]. Terpenes isolated from *G. lucidum* were found to have cytotoxic activity against several tumor cell lines. The alcohol extract of *G. lucidum* was shown that it inhibited cell proliferation in a dose and time dependent manner. Triterpene fraction of *G. lucidum* significantly inhibited human hepatoma cells. Oxygenated triterpenes isolated from *G. lucidum* spores also showed direct cytotoxicity *in vitro* on tumor cell lines [18]. *G. lucidum* triterpenes were shown to possess profound activity against leukemia, lymphoma, and multiple myeloma (Table 4).

A lot of scientific attention has been focused on *Ganoderma* polysaccharides which represent a structurally diverse class of biological macromolecules with a wide range of properties. Reports on pharmacological activity of *Ganoderma* polysaccharides mainly focus on antitumor effects[19]. The study on the anti-tumor effect of *G. lucidum* and their mechanism have been a matter of great interest. The antitumor effect of *G. lucidum* was mediated by cytokines released from activated T lymphocytes and macrophages. The currently available information suggests that intrinsic immunological mechanisms are involved in the antitumor activity of *G. lucidum*. Antitumor effect also has been demonstrated by inhibiting angiogenesis. The attribute of angiogenesis potential might be the direct inhibition of vascular endothelial proliferation or direct decrease of growth factor expression of tumor cells[20]. Clinical studies also showed that *G. lucidum* preparations exerted synergistic therapeutic effect when used in conjunction with radiation and chemotherapy and also reduced the side effects of these treatments (Table – 5).

It is generally accepted that the antitumor and anticancer effects of *Ganoderma* polysaccharides arise from the enhancement of body's immune system rather than direct cytotoxic effects. However, Reishi – *G. lucidum* is the most active cytotoxic to cancer cells among the large number of Basidomycetes species being tested. Several studies correlated the tumoricidal effects of Reishi with regulation of cancer cell cycling and signaling [21]. Recent preclinical studies undertaken in our laboratory showed that *G. lucidum* possessed significant protective effect against mammary adenocarcinoma and skin tumors [16].

Clinical observation assessment of 547 medium and late phase cancer patients treated with Chinese *G. lucidum* essence showed that the death rate of patients in long-term treatment was significantly low. A continuous 2-3 month active treatment with a daily dose of 4-6 g *G. lucidum* essence was proposed. Two randomized trials using *G. lucidum* polysaccharide (Ganopoly) was reported by Gao et al[21]. A total of 34 advanced patients of different cancer origin were recruited and administered 12 weeks with Reishi capsules at a dose of 1800mg, 80% of patients were found to have a series of immunological enhancements. In another study same capsules were administered to 68 lung cancer patients and over 65% of the patients were found to have improvements in quality of life. Based on available consolidated scientific evidence and clinical observations, Red mushroom – *G. lucidum* is beneficial to cancer patients [13].

Table 4: Major biological activities of triterpenes from *Ganoderma lucidum*

Name of the triterpene	Biological activity
Triterpenes	Hepatoprotective Anti-tumor Anti-angiogenic Hypotensive Hypocholestrolemic Anti-histaminic Platelet aggregation Anti-HIV
Ganoderic acids U, V, W, X, Y	Cytotoxic against hepatoma cells
Ganoderic acids A, B	Inhibition of farnesyl protein transferase
Ganoderic acid F	Prevention of invasion of metastatic cells *in vivo*
Ganoderic acid β	Inhibition of HIV-1 protease
Ganolucidic acid A	Inhibition of HIV-1 protease
Ganodermic acid S	Inhibition of Ca^{2+} mobilization in platelets
Ganoderic aldehyde A	Cytotoxic against hepatoma and nasopharyx carcinoma cells
Lucilaldehyde B, C	Cytotoxic against Lewis lung carcinoma (LLC), Mouse sarcoma Meth-A, sarcoma 180 and Breast cancer cells T-47D
Lucidimol A, B; Ganodermanondiol, Ganodermanontriol, Ganoderiol F	Cytotoxic against LLC and Meth-A cancer cells
Ganodermanol, Ganodermadiol	Cytotoxic against LLC, Meth-A, sarcoma 180, and T-47D cells
Ganodermanontriol, Ganoderiol F	HIV-1 inhibitors
Lucidimol B, Ganodermanondiol, Ganodermanontriol	Inhibition of HIV-1 protease

Table 5: Pharmacological properties of Polysaccharides isolated from
Ganoderma lucidum

β-glucans	Inhibition of tumor growth of sarcoma –180 ascites in mice
Polysaccharide extracts	Inhibition of tumor growth of Lewis lung carcinoma in mice
Sulfated D-glucans	Inhibition of tumor growth of Ehrlich ascites carcinoma in mice
Polysaccharide extracts	Prevention of cancer chemically induced colon cancers in rats
Polysaccharides (GLPS)	Mucosal healing chemically induced gastric ulcers in rats
Polysaccharide (GLP)	Hepatoprotective effects against hepatic inflammation in mice
Protein-bound polysaccharides	Hepatoprotective effects
Glycans (ganoderan A and B)	Hypoglycemic activity in alloxan-induced diabetic mice
Polysaccharides (Gl-PS)	Hypoglycemic activity in alloxan-induced diabetic mice
Polysaccharide extracts	Antihypertensive effects in rabbits and rats Antitumor effect through immunomodulation Antitumor effect through anti-angiogenesis Cell protection from free radicals
Preclinical studies	
Ganopoly (polysaccharide extracts)	Stimulation of immune system, increased plasma concentration of IL-2, IL-6, IFN-γ advanced-stage cancer of lung, colon, breast, liver, prostate, bladder and brain

Hepatoprotective activity

G. lucidum has been reported to be widely used for the treatment of chronic hepatopathy of various etiologies. Several animal studies have demonstrated that Red mushroom extract exhibits protective effect against liver injury induced by toxic chemicals. Studies carried in our laboratory showed that hepatotoxicity induced by carbon tetrachloride and alcohol was effectively ameliorated by alcoholic extract of *G. lucidum* [22,23]. The polysaccharide fractions and triterpenes have shown protective effects on liver in animal and human studies. Ganoderic acids R and S from the cultured mycelia of *G. lucidum* showed strong anti-hepatotoxic activity in galactosamine-induced cytotoxicity to cultured rat hepatocytes. Reishi and its extracts can prevent liver damage induced by CCl_4 and galactosamine in rats [24]. Triterpenoids isolated from *G. lucidum* showed significant protection against immunological liver damage in mice *in vivo* and *in vitro*.

Hepatoprotective effect of Reishi – *G. lucidum* polysaccharides were evaluated in mouse model and the treatment with polysaccharides diminished histological changes of injury and inflammatory infiltration of lymphocytes. Immunohistochemistry showed that polysaccharides inhibited expression of inducible nitric oxide synthase suggesting the protective effect was mediated by the inhibition of nitric oxide (NO). Protein-bound polysaccharides from Reishi- *G. lucidum* reduced serum aspartate transaminase (AST), alanine transaminase (ALT), alkaline phosphatase (ALP) and total bilirubin in rats with cirrhosis induced by biliary obstruction.

Chinese doctors demonstrated that polysaccharide- containing preparations of *G. lucidum* had curative effect on patients with chronic hepatitis B. Following treatment with Ganopoly for a period of 6 months, aminotransferase levels of 33% patients returned to normal, 13 % patients cleared hepatitis B surface antigen from serum whereas none of the controls were normal. The drug was well tolerated and the study revealed that Ganopoly is significantly active against HBV[25]. In another study, 367 patients with chronic hepatic diseases such as hepatitis and liver cirrhosis were treated with *G. lucidum* extract as tea with good results. Most patients showed subjective symptomatic improvement.

Antidiabetic effect

Diabetes mellitus which includes diabetes type 1 and type 2 patients, is a disease caused by abnormal glucose metabolism. The disease is affecting 250 million people worldwide. More researches are focused on effective and safer treatment modalities for peripheral insulin resistance in case of type 2 diabetes patients. Water extract of Reishi reduced the increase in blood glucose and blood insulin level in oral glucose test. Following the administration of mushroom extract the increase in blood glucose was inhibited without raising blood insulin levels. Animal studies have demonstrated that polysaccharides isolated from Reishi have potential hypoglycemic and hypolipidemic activities. Glycans (ganoderan B and D) showed significant hypoglycemic activity in mice. Ganopoly was studied for its effect on carbohydrate and lipid metabolism and the results showed that Ganopoly possessed therapeutic effect on diabetic rats; it may promote insulin secretion.

In clinical studies, Ganoderan A and B, glucans isolated from *G. lucidum* fruit bodies ameliorated the symptoms of diabetes. In one, study 71 type 2 diabetic patients were treated with Ganopoly (1800 mg) three times daily for 12 weeks. Fasting and stimulated glycosylated hemoglobin (HbA1c), plasma glucose, insulin, and C-peptide were monitored at predetermined intervals. Treatment with Ganopoly significantly decreased the mean HbA1c from 8.4% at base line to 7.6% at 12 weeks. The studies showed that Ganopoly is well tolerated. This demonstrated that Ganopoly is efficacious and safe in antidiabetic treatment in patients [26].

For a clinical investigation using Green Valley Lingzhi capsules, 130 patients suffering type 2 diabetes mellitus were recruited. After 2 months of treatment with the formulation, showed synergistic effect in hypoglycemic action combined with regular hypoglycemic drugs, significantly decreased clinical symptoms, compared to control group treated only with regular hypoglycemic drugs.

Cardioprotective effect

Cardiovascular diseases are one of the major causes of death in the world particularly in developed industrial countries. Hypercholestrolemia and associated cardiovascular diseases remains the leading cause of morbidity and mortality. Therapeutic agents that can protect the cardiac myocites by acting as a cellular scavenger of ROS, hypolipidemic agents and protector of innate antioxidant defense system of mitochondria without inducing any toxicity can be useful as cardioprotective agent. Recent studies in our laboratory demonstrated that *G. lucidum* significantly protected mitochondria by preventing the decline of antioxidant status and membrane potential by directly scavenging the free radicals [27].

ROS and increased levels of blood lipids are key elements in the pathogenesis of atherosclerosis. The control of blood lipids especially cholesterol, is important for reducing the risk of the development or progression of atherosclerosis. Myocardial injury shares a major etiological mechanism for aging in that both involve an increase in ROS generation and oxidative stress. The aging heart undergoes significant functional and structural alterations leading to atrophy and compensatory hypertrophy followed by myocardial fibrosis. In addition there is an age related decline in the capacity to withstand stress, such as ischemia, reperfusion. In its severe form cardiac decay results in congestive heart failure, one of the leading causes of death in people over the age of 65. Ganopoly has been found to enhance antioxidation and reduce lipid peroxidation in rats with hyperlipidemia .Other reports also reveal that triterpenoids isolated from *G. lucidum* possessed the ability to inhibit the biosynthesis of cholesterol and angiotensin converting enzyme.

Ganoderma has been reported to decrease high blood pressure and this has been attributed to the ganoderic acids. *Ganoderma* triterpenes also has been reported to inhibit platelet aggregation. The effect of Reishi on cardiovascular system has been investigated. The decrease in high blood pressure was reported to be attributed to ganoderic acid. Triterpenes from Reishi that inhibit ACE have also been described. Inhibition of cholesterol biosynthesis, enhanced antioxidative activity, decreased platelet aggregation [4] and reduced lipid peroxidation have been demonstrated in animals and *in vivo* experiments. Adenosine content in the crude extract of *Ganoderma* has been shown to inhibit platelet aggregation. Reishi can counteract the lipid peroxidation and its therapeutic mechanism for atherosclerosis may be related to the counteraction of lipid peroxidation and enhancement of endogenous antioxidat activity[15].

Other reports previously indicated that triterpenoids isolated from *G. lucidum* possessed the ability to inhibit the biosynthesis of cholesterol and contributed to atherosclerosis protection by inhibition of ACE or platelet aggregation. Reishi may affect cholesterol synthesis and the oxygenated sterol from *G. lucidum* can inhibit cholesterol biosynthesis via conversion of acetate or mevalonate as a precursor of cholesterol. Some triterpenes from *G. lucidum,* such as ganoderic acid F, contributes to atherosclerosis protection by inhibiting ACE or of platelet aggregation. At high concentration of ganoderic acid S, platelet aggregation occurred while at low concentration the aggregation was inhibited. The aggregation was concentration and time depended.

The water extract of mycelia of *G. lucidum* was also assessed for cardiovascular activity in rabbits and rats. Although the mushroom decreased systolic and diastolic blood pressure, there was no difference in the heart rate of the tested animals. It has recently been found that nucleosides which contain adenosine and guanosine, found in the water/alcohol extract of Reishi possess platelet aggregation inhibition action (antithrombotic activity). A water extract of mycelium administered to rats and rabbits (3-30 mg/kg i.v.) produced significant hypotensive effects, an activity suggested secondary to primary effect that suppresses sympathetic outflow of the central nervous system. The powdered mycelium of Reishi lowered systolic blood pressure without causing significant difference in heart rate of spontaneously hypertensive rats for 4 weeks.

In a double –blind placebo-controlled clinical study of *G. lucidum* in 54 patients with primary stage II hypertension who had not responded to previous drug treatment were recruited. The group which was administered *G. lucidum* extract tablets (2 tablets b i d or 200 mg/day), systemic blood pressure improved in 82.5%, with capillary and arterial blood pressure showing significant improvements in 14 days. No change of any significance was found in placebo. In treating hypertension, *G. lucidum* was shown highly effective in a very large number of treated cases. In more successful cases, blood pressure was back to normal within 2 months, in some cases within 2 weeks.

A peptidoglycan (MW 100,000) having mild hypotensive effect on rats has been isolated from hot water extract of Reishi. According one report, blood pressure about half a number of patients with hypertension was reduced when Reishi extract was administered. It has been reported that a hypotensive related ACE was inhibited by ganoderic acids B,D,F,H,R,S and Y, ganoderal A and ganoderol A and B.

Dose and toxicity

Practitioners of TCM recommend 0.5 to 1 g extract of Reishi for a normal person as health tonic, 2 to 5 g extract for chronic illness and up to 15 g extract daily for serious illness. The Chinese pharmacopoeia recommends 6 to 12 weeks were used in clinical trials without toxicity. Most common dose of Reishi is 250 mg (standard extract) 4 times a day with food.

Almost all toxicity studies with *G. lucidum* in animal models showed low or no detectable toxicity. There are no reported data on adverse effects on lone term administration of *G. lucidum* and its derivatives. Aqueous extract of Reishi administered to mice orally at a dose of 5g/kg body weight for 30 days produced no change in body weight, organ weight or hematological parameters. The polysaccharides fraction at the same dose also produced no side effects. All these studies show that Reishi-*G. lucidum* is absolutely safe.

Summary and future prospects

Reishi- *G. lucidum* has been used as medical remedy in TCM and in many Asian countries during the past two thousand years. The medicinal effects of this mushroom are fascinating and it is considered as a medical wonder. The mushroom is thought to preserve human vitality and to promote longevity. Reishi has been used to treat various diseases such as allergy, arthritis, bronchitis, gastric ulcer, hyperglycemia, hypertension, chronic hepatitis, insomnia, nephritis, neurasthenia, scleroderma, inflammation and cancer. In addition, this mushroom has been reported to have various pharmacological effects which include, immunostimulatory, analgesic, antiaging, antiatherosclerotic, antibacterial, antinociceptive, antioxidant, inhibition of platelet aggregation, antithrombotic, antiviral, chemopreventive, hepatoprotective, hypoglycemic, hypolipidemic, antiulcer, nephroprotective and radioprotective effects.[11, 18,28-33]

Reishi – *G. lucidum* contains a wide variety of bioactive molecules which included triterpenoids, polysaccharides, nucleotides, sterols, steroids, fatty acids, proteins/ peptides, and several trace elements. However, triterpenoids and polysaccharides are the major bioactive compounds. Because of its demonstrated health benefits, Reishi has gained great popularity in recent years not only in China and Japan but also in North America and other parts of the world. The reason of its international attention as a valuable Chinese medicine is because of its wide variety of biological activities. *Ganoderma* products derived from the fruiting bodies and mycelia have attracted great deal of attention during the last decade. Several products have undergone clinical trails and became available commercially as syrup, injection, tablet, tincture and mushroom powder. Fruiting body powder capsules or tablets of Reishi have been demonstrated to possess preventive and curative effects on diseases such as cardiovascular diseases, hypertension, hepatitis, diabetes, tumor and cancer. Based on consolidated scientific evidence available so far, Reishi has significant health benefits and would soon become a new therapeutic agent for treatment of a wide range of diseases.

References

1. Chang, S.T. and Miles, P., Mushrooms, Cultivation, Nutritional value, Medicinal effects and Environmental impact, CRC Press, New York, 2004.

2. Wasser,S.P., Medicinal mushrooms science : History current status, future trends, and unsolved problems, *Int. J. Med. Mushr.*,2010,**12**,1-16.

3. Jong S and Birmingham, J.M., Medical benefits of the mushrooms *Ganoderma. Adv. Appl. Microbiol.*, 1992, **37** 101-134.

4. Zhou, X., Lin, J., Yin, Y., Zhao, J., Sun, X. and Tang, K., *Ganodermataceae*: Natural products and their related pharmacological functions, *The Am. J. Chinese Medicine,* 2007,**35**, 559-574.

5. Wasser S.P., Reishi or Ling Zhi (*Ganoderma lucidum*) Encylo . Diet. Supplements, 2005 603 – 622.

6. Bao, X.F ., Liu .C.P., Fang J.N. and Li. X. Y., Structural and immunological studies of a major polysaccharide from spores of *Ganoderma lucidum* (Fr.)P. Karst. *Carbohydr. Res*, 2001 332 **67**-74.

7. Peterson R.R.M., *Ganoderma* – A therapeutic fungal biofactory, *Phytochemistry* 2006,**67,**1985-2001.

8. Kim,H.W; Kim,B.K. Biomedical triterpenoids *Ganoderma lucidum* (Curt.:Fr.) P.Karst (Aphyllophoromycetidae) *Int .J. Med . Mushr,* 1999 **1,** 121- 138.

9. Jones S. and Janardhanan, K.K., Antioxidant and antitumor activity of *Ganoderma lucidum* (Curt:Fr.) P.karst. – Reshi (Aphyllophoromycetidae) from South India *Int J. Med. Mushr,*2000, **2,**195-200.

10. Sheena, N., Ajith, T.A and Janardhanan, K.K., Prevention of nephrotoxicity induced by the anticancer drug cisplatin, using *Ganoderma lucidum*, a medicinal mushroom occurring in South India, *Curr. Sci.* 2003, **85,** 478-482.

11. Sheena, N., Ajith, T.A. and Janardhanan, K.K., Anti – inflammatory and anti – nociceptive activities of *Ganoderma lucidum* occurring in South India, *Pharmaceutical Biol,*2003, **41,** p301-304.

12. Boh, B., Berovic M., Zhang J . and Zhi-Bin L., *Ganoderma lucidum* and its pharmaceutically active compounds *Biotech . Annual Review* 2007 **13,** 265- 301.

13. Yuen J.W.M. and Gohel M.D., Anticancer effect of *Ganoderma lucidum* : A review of scientific evidence, *Nutr Cancer,*2005,**53,**11-17.

14. Smith, J., Rowan, N. and Sullivan, R., Medicinal mushrooms: Their Therapeutic properties and Current Medical Usage With Special Emphasis on Cancer Treatment; Special Report Commissioned by cancer Research UK; The University of Strathclyde in Glasgow, 2001 256.

15. Sheena, N., Ajith, T.A. and Janardhanan, K.K., Protective effect of methanolic extract of *Ganoderma lucidum* P.Karst. Reshi from South India against doxorubicin –induced cardiotoxicity in rats, *Orient Pharmacol Exp.Med* 2005 **5,**62-68.

16. Lakshmi, B., Sheena, N. and Janardhanan, K.K., Prevention of mammary adenocarcinoma and skin tumour by *Ganoderma lucidum* a medicinal mushroom occurring in south India, *Curr. Sci.,*2009, **97,** 1658 1663.

17. Joseph,S., Sabulal, B., Antony, A.R. and Janardhanan, K.K., Antitumor and anti-inflammatory activities of polysaccharides isolated from *Ganoderma lucidum,* *Acta .Pharma,*2011,**61,**335-342.

18. Sliva, D., Cellular and physiological effects of *Ganoderma lucidum* (Reishi).*Mini –Reviews in Med. Chem,* 2004 **4** 873- 879 .

19. Wasser, S.P., Medicinal mushrooms as a source of antitumor and immunomodulatory polysaccharides . *Appl .Microbiol. Biotechnol,* 2002, **60,** 258- 274.

20. Zhang, N.M., Zhang, L., Isolation, Purification and Pharmacological actions and application to functional food of *Ganoderma lucidum* polysaccharide. *Food Res. Dev,* 2005 **26** 118-120.

21. Gao, Y., Zhou, S., Jiang, W., Huang, M. and Dai, X., Effects of ganopoly (a *Ganoderma lucidum* polysaccharide extract) on the immune functions in advanced –stage cancer patients. *Immunol Invest* 2003, **32** 201-215.

22. Sheena, N., *Studies on the therapeutic potential of Ganoderma lucidum P.Karst –
    Reishi, occurring in Kerala,* Ph.D Thesis, Calicut University, 2005.

23. Lakshmi, B., Ajith, T. A., Jose, N. and Janardhanan, K.K., Antimutagenic activity
    of methanolic extract of *Ganoderma lucidum* and its effect on hepatic damage
    caused by benzo[*a*] pyrene, *J. Ethanopharmacology* 2006,**107**,297-303.

24. Sudheesh, N.P., Ajith, T.A., Mathew, J., Nima N. and Janardhanan, K.K.,
    Ganoderma lucidum protects liver mitochondrial oxidative stress and improves
    the activity of electron transport chain in carbon tetra chloride intoxicated rats,
    Hepatol Res 2012,**42**,181-191.

25. Gao, Y., Zhou, S., Chen, G., Dai, X., Ye, J. and Gao, H., A Phase I/II study of a
    Ganoderma lucidum (Curt : Fr) P. Karst . (Ling Zhi Reishi mushroom) extract
    in patients with chronic hepatitis B. *Int. J. Med . Mushrooms,* 2002, **4** 321-327.

26. Gao, Y., Lan J. Dai, X., Ye J. and Zhou, S., A Phase I/II study of Ling Zhi mushroom
    Ganoderma lucidum (W.Curt : Fr) P. Karst . Lloyd (Aphyllophoromycetidae)
    extract in patients with type II diabetes mellitus . *Int. J. Med . Mushrooms* 2004,
    6 3-9 .

27. Sudheesh N.P., Ajith, T.A. and Janardhanan, K.K., *Ganoderma lucidum* ameliorate
    mitochondrial damage in isoproterenol –induced myocardial infarction in
    rats by enhancing the activities of TCA cycle enzymes and respiratory chain
    complexes, *Int. J. Cardiol,* 2013, **165**, 117-125.

28. Pillai T,G., Salvi,V.P., Maurya D.K., Nair, C.K.K. and Janardhanan, K.K,.
    Prevention of radiation–induced damages by aqueous extract of *Ganoderma
    lucidum* occurring in southern parts of India *Curr. Sci.,*2006, **91**,341-344.

29. Ajith, T.A., Sudheesh, N.P., Roshny, D., Abishek,G. and Janardhanan, K.K.,
    Effect of *Ganoderma lucidum* on the activities of mitochondrial dehydrogenases
    and complex I and II of electron transport chain in the brain of aged rats, *Exp.
    Gerontology* 2009, **44**, 219-223.

30. Sudheesh,N.P., Ajith, T.A., Ramnath,V. and Janardhanan, K.K., Therapeutic
    potential of *Ganoderma lucidum* (Fr.)P. Karst .against the declined antioxidant
    status in the mitochondria of post-mitotic tissues of aged mice. *Clinical Nutrition,*
    2010, **29**, 406-412.

31. Rony, K.A., Mathew, J., Neenu, P.P. and Janardhanan, K.K., *Ganoderma
    lucidum*(Fr.) P.Karst occurring in South India attenuates gastric ulceration in
    rats, *Indian J. Natural Products and Resources* 2011, **2**, p19-27.

32. Smina T.P., De, S., Devasagayam, T.P.A., Adhikari .S. and Janardhanan, K.K.,
    Ganoderma lucidum total triterpenes prevent radiation–induced DNA damage
    and apoptosis in splenic lymphocytes in vitro, *Mutation Res* 2011,**726**,188-194.

33. Smina,T.P.,Mathew, J., Janardhanan, K.K. and Devasagayam,T.P.A., Antioxidant
    activity and toxicity profile of total triterpenes isolated from *Ganoderma lucidum*
    (Fr.)P.Karst occurring in South India, *Env. Tox.Pharmacology,*2011,**32**,438-446.

4

Medicinal and chemical aspects of some wild mushrooms occurring in Kashmir-India

Abdul. S. Shawl* and Gulzar. A. Bhat

CSIR-Indian Institute of Integrative Medicine (Formerly Regional Research Laboratory), Sanatnagar, Srinagar-190 005 –India.
* E-mail: asshawl@gmail.com

Mushrooms are known to humankind from times immemorial because of the fact of their nutritional and medicinal importance. Mushrooms have a long history of use in traditional medicine and have shown antitumor, immunomodulatory, antioxidant, hepatoprotective, antidiabetic, antiviral, properties etc. Total trade in medicinal mushrooms in the form of dietary supplements is well over 18 billion USD. Out of six dietary supplements consumed in USA, at least one contains a mushroom as it's ingredient in one form or the other. A number of bioactive molecules have been isolated and characterized in the form of polysaccharides (hetero β-glucans with 1,3 linkages in the main chain and β-1,6 linkages in the branched chain), polysaccharide protein complexes, triterpenoids, sesquiterpenes, sterols, flavour compounds, coumarins, alkaloids, fatty acids, phenolics etc. Kashmir valley due to its unique landscape and climatic condition is a rich repository of mushroom species. As part of institute's programme on resourcing of biomolecules from medicinal plants including mushrooms, a number of mushroom species were collected from different regions of Kashmir, properly identified on the basis of morphological, reproductive and other recorded data. Different extracts were prepared as per standard operating procedures following traditional as well as modified extraction methods. *Russula brevipes* showed promising anticancer activity, while as *Lactarius controversus* and *Phellinus pomaceus* showed hepatoprotective activity. The present paper discuss an overview of global status of mushrooms as sources of anticancer and immunomodulating adjuvants, survey and collection of some mushroom species with medicinal and chemical aspects, modern methods of extraction, isolation and bio evaluation studies.

Introduction

Mushroom is regarded as macrofungus with a distinctive fruiting body that can be either epigeous or hypogeous and large enough to be seen with naked eye and to be picked by hand[1]. Mushrooms are somehow fascinating due to their unusual characteristics, belonging to the kingdom fungi, which constitutes the most diverse group of organism on this biosphere. For millennia, mushrooms have been valued by human-kind as an edible and medical resource[2]. Medicinal mushrooms (MMs) are a big source of polysaccharides, polysaccharide-protein complexes mostly belonging to β-glucans having 1→3 linkage in the main chain and 1→6 linkage in the branches. Various bioactive compounds prominent in mushrooms are low molecular weight compounds cerebrosides, isoflavonoids, catechols, amines, triglycerols, sesquiterpenes, steroids, organic germanium and quinines[3]. A vast medicinal function are thought to be produced by MMs and fungi, including antitumor, immunomodulating, antioxidant, radical scavenging, cardiovascular, antihypercholesterolemic, antiviral, antibacterial, antiparasitic, antifungal, detoxification, hepatoprotective, and antidiabetic effects [4]. Lentinan, Krestin (PSK) and Active hexose correlated compound (AHCC) are licensed pharmaceuticals in China, Japan, Korea and other countries. Several of the mushroom polysaccharide compounds have proceeded through Phase I, II, and III clinical trials and are used extensively and successfully in Asia to treat various cancers and other diseases[5]. In view of usage of combination therapies presently for cancer treatment and other diseases, MMs have became more important and are popularly known as biological response modifiers [6] because of the fact of activating different immune response. Modern scientific studies on MMs have expanded exponentially during the last two decades and scientific explanation to show how compounds derived from mushrooms function in humans are increasingly being established (Fig. 1)[7].

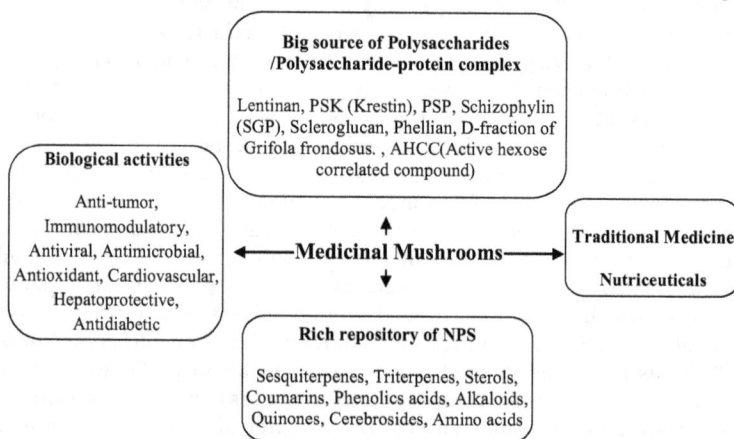

Big source of Polysaccharides /Polysaccharide-protein complex

Lentinan, PSK (Krestin), PSP, Schizophylin (SGP), Scleroglucan, Phellian, D-fraction of Grifola frondosus. , AHCC(Active hexose correlated compound)

Biological activities

Anti-tumor, Immunomodulatory, Antiviral, Antimicrobial, Antioxidant, Cardiovascular, Hepatoprotective, Antidiabetic

Medicinal Mushrooms

Traditional Medicine

Nutriceuticals

Rich repository of NPS

Sesquiterpenes, Triterpenes, Sterols, Coumarins, Phenolics acids, Alkaloids, Quinones, Cerebrosides, Amino acids

Fig.1: General view of Medicinal Mushrooms

Recent advances in extraction technologies and structural elucidation

Mushrooms are attracting more and more attention for its medical foods and antitumor value. Enormous studies have been made during the last two decades in the extraction technologies of polysaccharides.[8] Modern separation technology is recommended to produce high quality mushroom functional foods. A series of mushroom functional foods can be prepared from fresh or dried mushrooms using new technology such as supercritical fluid extraction, high vacuum and low temperature molecular distillation, extrusion, ultrafiltration. Ultrasonic assisted extraction (UAE) optimized by response surface methodology is the modern method for extraction of polysaccharides from mushrooms. These kinds of extracts maintain the bioactive components, maximal biological activities, and compete efficiently with market terms. Also there is the full utilization of the materials present in the stock extract. This method has been applied for polysaccharide extraction from dried longan pulp[9] and extraction of water soluble polysaccharides from submerged cultured mycelia of *Boletus edulis*.[10] A novel process for nanoparticle extraction of the β-D-glucans from *Sparassis crispa*, and Phellin, the β-D-glucan from *Phellinus linteus*, has been investigated using insoluble tungsten carbide as a model for nanotechnology[11]. Mizumo developed an excellent extraction method for fractionation and purification of polysaccharides (Fig. 2)[5, 8]. Extracted polysaccharides can be further purified using a combination of techniques such as ethanol precipitation (excludes the impurities from the polysaccharides), fractional precipitation (acidic precipitation with acetic acid), ion-exchange chromatography (separation of acidic and neutral polysaccharides), gel filtration and affinity chromatography (separation of neutral polysaccharides into α-glucans (adsorbed fraction) and β-glucans (non-adsorbed fraction). On the basis of total acid hydrolysis, methylation analysis, per iodate oxidation, NMR experiments (^{1}H, ^{13}C, DEPT-135, DQF-COSY, HMQC, HMBC, NOESY, ROESY and TOCSY), and MALDI-TOF (Mass spectrometry) the structures of these polysaccharides are elucidated.[12] Similarly high performance ion exchange chromatography and application of DOSY-NMR (by Open source software) coupled with GC-MS spectrometry have been quite useful in the structure elucidation of mushroom metabolites.[13]

Collection

Collection of the fresh mushrooms has been achieved from target sites, mostly from Gulmarg (2500 m above sea level), Tangmarg (1900 m above sea level), Pahalgam (2600 m above sea level), Sonamarg (2700 m above sea level) and Srinagar. The collected mushroom samples were identified on the basis of morphological, reproductive characters and comparing with standard field guides by Largent (1973), mushroom herbaria of IIIM and National Centre for Mushroom, Solan, Himachal Pradesh, India. The list of the mushrooms following standard methods of collection from Kashmir valley are shown in Table 1.

Table 1: Collection of various Mushrooms in Kashmir valle

S. No.	Name of the species	Place of collection
1.	*Amanita citrina*	Pahalgam
2.	*Amanita vaginata*	Harwan
3.	*Clavulinopsis corniculata*	Gulmarg
4.	*Coprinus atramentarius*	Shallabugh /Ganderbal
5.	*Coprinus comatus*	Shallabugh/Ganderbal
6.	*Fomes fomentarius*	Majid bagh/Narbal
7.	*Ganoderma lucidum*	Gulmarg/Sanatnagar
8.	*Ganoderma tsugae*	Gulmarg
9.	*Gyromitra esculenta*	Shallabugh/Ganderbal
10.	*Lactarius controversus*	Nowgam /Kanipora
11.	*Lentinus tigrinus*	Ganderbal
12.	*Morchella conica*	Ganderbal
13.	*Morchella esculenta*	Ganderbal
14.	*Phellinus pomaceus*	Majid Bagh/Gangbugh
15.	*Pholiota destruens*	Kanipora/Gulmarg
16.	*Pleurotus flabellatus*	Sanatnagar
17.	*Pleurotus sajor-caju*	Gulmarg
18.	*Polyporus brumalis*	Gulmarg
19.	*Russula brevipes*	Gulmarg
20.	*Trametes versicolor*	Bandipora

Extraction methods

Polysaccharides from different mushrooms were separated by successive aqueous extractions with boiling water, with NaOH aqueous solution (1 M, 100 °C), and 80% ethanol (EtOH). Extraction with 80% EtOH yielded low molecular substances, extraction with hot water yielded water-soluble polysaccharides, and extraction with alkali yielded water-insoluble polysaccharides (Fig. 2).

In our institute, we usually follow the standard operating procedure for extraction such as ethanol (100%), water (100%) and ethanol-Water (50:50) (Fig. 3). Sufficient quantity of solvent is added to submerge the mushroom material. After standing for about 16 hrs, the percolate is collected and filtered if required. This process of extraction is repeated four times. The combined extract is evaporated to dryness under reduced pressure at below 50 °C using rotavapor. The final drying may be done in a vacuum desiccators or lyophilizer depending on the nature of extract. The dried extract is scraped off and transferred to a tarred wide mouth glass container of appropriate size. Nitrogen is blown in the container before capping. The container is weighed to calculate the quantity of extract obtained. This forms the stock extract of the mushroom and is labeled. Fash chromatography is used to get target fractions. Labeled extract is onwards used for isolation of bioactive constituents and for different biological activities. For characterization of structure, molecular weight and molecular weight distribution of oligo/polysaccharides MALDI-TOF-TOF Mass spectrometer was used.

Bioevaluation

Mushroom species (fresh as well as dried) collected were subjected to solvent extraction. Bioevaluation of the extracts for antibacterial, antifungal, immunomodulatory, hepatoprotective and anticancer activities were carried out (Table 2, 3, 4 & 5). The samples showing MIC of 2500 µg/ml and 500 µg/ml or below were considered active for

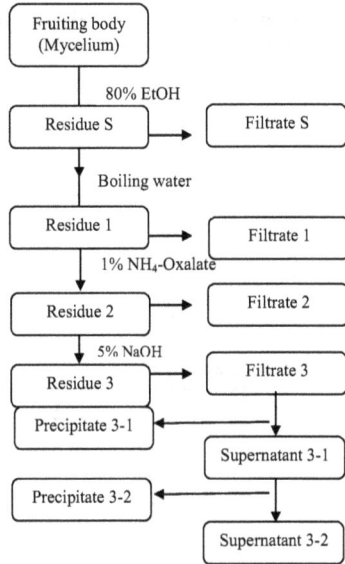

Fig 2: Fractional preparation of polysaccharides from mushrooms

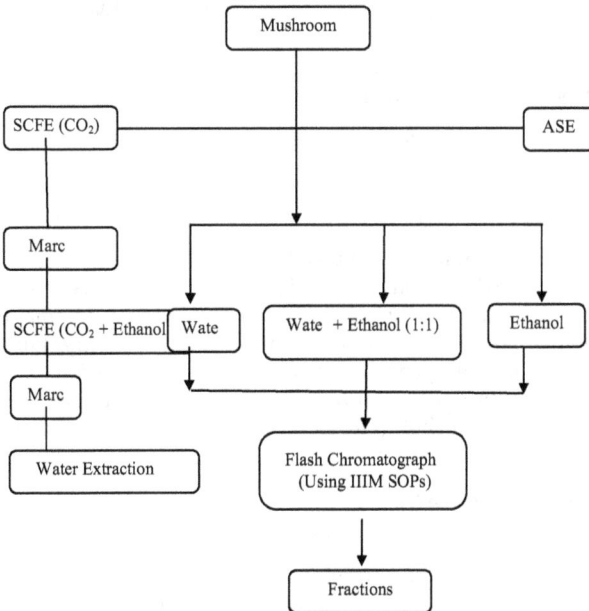

Fig. 3: Extraction procedures of mushrooms

antibacterial and antifungal activities respectively[14]. In immunomodulatory assay, *in vivo* humoral (antibody titre) and cellular (DTH) response were determined at 100 mg/kg in mice. Levamisole showed 138 and 140% activity against these parameters respectively. The samples showing activity better than Levamisole on either of the parameter were considered active. *In vitro* flow cytometric analysis of TNF-α were estimated in human blood neutrophils using PE-labelled TNF-α monoclonal antibodies by flowcytometry.[15] The samples showing 35% or more decrease in TNF-α levels at 100 µg/ml were considered immunosuppressive. Hepatoprotective properties of the mushroom extracts (100 and 200 mg/kg/day, p.o. for 7 days) were evaluated on paracetamol (2 g/kg, p.o)[16] induced hepatotoxicity.

Table 2: Antibacterial activity of *Rassula brevipes*

S.No.	Sample Code	MIC in µg/ml			
		ATCC-29213	MRSA-15187	ATCC *E. coli*	ATCC *P. aeroginosa*
1	Rassula brevipes A001	2000	200	>4000	>4000
2	A002	4000	4000	>4000	>4000
3	H	4000	4000	>4000	>4000
4	A003	>4000	>4000	>4000	>4000
Standard	Ciprofloxacin	0.25	8.0	<0.035	0.25

A001=Ethanol extract; A002=Water extract; A003=Ethanol-Water (1:1);

H=Hexane extract

Wild mushrooms of Kashmir

The different wild mushrooms growing in Kashmir are shown below.

Amanita citrina

Common name: False death cap, Citron Amanita

Amanita citrina, often referred to as the False Death cap is an inedible fungus, although not reported to be seriously toxic (Figure 4) . *A. citrina* has been shown to possess cytotoxic and antibiotic activities. Bioactive molecules in amanita citrina include bufotenine-N-oxide, serotonin, 5-methoxy-N-dimethyltryptamine, N,N-dimethyltryptamine, monoterpenes (α-pinene, limonene).[18-21] Various extracts of *A. citrina* didn't show any promising antifungal, anticancer and immunomodulatory activity. Water extract and alcohol:water extract showed moderate hepatoprotective activity (Table 6).

Clavulinopsis corniculata

Common name: Meadow coral; Cow coral fungus

Clavulinopsis is a genus of coral fungi occur solitary or in clumps on soil among decaying forest liter. *Clavulinopsis corniculata* distinctive among corals and is saprobic, growing alone under hardwoods or in grassy areas (Figure 5). In our lab, different

Fig. 4: *Amanita citrina* (http://www.toadstool.ru)

Fig. 5: *Clavulinops corniculata* (http://www.first-nature.com)

Fig. 6: *Coprinus atramentarius* (http://www.fungikingdom.net)

Fig. 7: *Inonotus hispidus* (http://mycoweb.narod.ru)

Fig. 8: *Lactarius controversus* (Laurent Francini)

Fig. 9: *Phellinus pomaceus* (http://www.mykopedia.org)

Fig. 10: *Russula brevipes* (http://w2.cegepsi.ca)

Fig. 11: *Trametes versicolor* (http://en.wikipedia.org)

Table 3: Antifungal activity

S.No.	Sample Code	MIC in µg/ml			
		Candida albicana (ATCC-90028)	*Candida albicana* (V-O-191)	*Aspergillus fumigata*	*Aspergillus niger* (ATCC-16409)
1	*Rassula brevipes* A001	>4000	>4000	>4000	>4000
2	A002	>4000	>4000	>4000	>4000
3	H	>4000	>4000	>4000	>4000
4	A003	>4000	>4000	>4000	>4000
5	*Clavulinopsis corniculata* H	>4000	>4000	>4000	>4000
6	A001	>4000	>4000	>4000	>4000
7	A002	>4000	>4000	>4000	>4000
8	A003	>4000	>4000	>4000	>4000
9	*Amanita citrina* A002	>4000	>4000	>4000	>4000
10	A003	>4000	>4000	>4000	>4000
Standard	(Amphoterrin-B)	0.25	0.25	0.5	0.25

A001=Ethanol extract; A002=Water extract; A003=Ethanol-Water (1:1); H=Hexane extract

Table 4: Immunomodulatory activity

Treatment	Dose (mg /Kg) P.O	Humoral Immune Response (*in vivo*)		Results
		Antibody titre Mean ± S. E.	Immuno-modulatory activity (%)	
Control	-	5.66±0.21	-	
Cyclophosphamide	250	4.00±0.00	29.32	
Clavulinopsis corniculata H	100	3.16±0.16	50.60	active
A001,2,3	100	3.50±0.22	30.12	
Amanita citrina A002,3	100	4.50±0.22	30.13	
Levamisole	2.5	6.33±0.21	140.36	

Table 5: Anticancer activity

Cell line Type				Lung	Colon	CNS
CCL CODE	Sample Code	Conc. µg/ml	A-549	HOP-62	HCT-15	SF-295
				Growth inhibition (%)		
M-1839	*Clavulinopsis corniculata* H	100	58	47	56	28
M-1840	A001	100	61	47	53	28
M-1841	A002	100	42	13	9	9
M-1842	A003	100	39	0	13	0
M-1843	*Amanita citrina* A002	100	31	8	7	0
M-1844	A003	100	33	4	0	0
	5-Fu	2×10^{-5} M	-	-	30	-
	Paclitaxel	2×10^{-5} M	15	11	-	-
	Adriamycin	2×10^{-5} M	-	-	-	42

A001=Ethanol extract; A002=Water extract; A003=Ethanol-Water (1:1); H=Hexane extract

Cell line type			liver		Neuro blastoma	colon
CCL CODE	Sample code	Conc. µg/ml	A-549	HOP-62	HCT-15	SF-295
				Growth inhibition (%)		
M-1824	*Rassula brevipes* A001	100	4	0	2	28
M-1825	A002	100	15	9	38	7
M-1826	H	100	19	0	33	21
M-1827	A003	100	17	0	42	19
	Mito-C	1×10^{-5} M	82	87	-	86
	Adriamycin	1×10^{-5} M	79	89	-	22
	5-Fu	2×10^{-5} M	-	-	27	15

Cell line type			lung		Neuro blastoma	prostate
CCL CODE	Sample code	Conc. µg/ml	A-549	IMR-32	DU-145	PC-3
			Growth inhibition (%)			
10289	*Tremetus versicolor* A001	100	22	58	65	13
10290	A002	100	19	28	55	13
10291	A003	100	21	12	43	15
	Mito-c	1x 10^{-5} M	84	67	72	12
	Paclitaxal	1x 10^{-5} M	27	47	70	22
	Adriamycin	2x 10^{-5} M	39	82	70	32

A001=Ethanol extract; A002=Water extract; A003=Ethanol-Water (1:1); H=Hexane extract

Table 6: Hepatoprotective activity

Sample code	% Hepatoprotective effect
Clavulinopsis corniculata	
H	60.6
A001	66.2
A002	53.1
A003	47.5
Amanita citrina	
A002	43.7
A003	40.0
Phellinus pomaceus	
A001	76
A002	74
A003	67
Lactarius controversus	
A001	86
A002	96
A003	64
Silymarin (Standard)	75.6

A001=Ethanol extract; A002=Water extract; A003=Ethanol-Water (1:1); H=Hexane extract

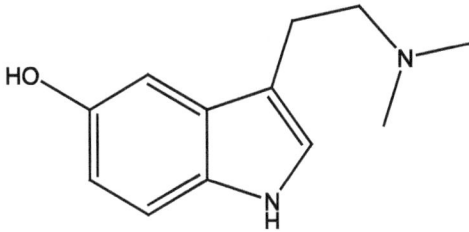

Bufotenine

extracts of this species were evaluated for different biological activities. Different extracts of this species show moderate hepatoprotective activity but no antifungal activity. Hexane and ethanol extracts show promising activity against A-549(lung cancer) and HCT-15 (colon cancer) cell lines. Hexane extract of *Clavulinopsis corniculata* showed good immunomodulatory activity.

Coprinus atramentarius

Common names: Inky cap, Tippler's bane

Local name: Vangan haddur

Coprinus atramentarius is a common edible mushroom in Kashmir (Figure 6). Extracts from the fruit bodies of *C. atramentarius* has been shown to inhibited the growth of Sarcoma 180 and Ehrlich solid cancers by 100%.[22] The aqueous extract of Tippler's bane reduce the mycelial growth and inhibit sporulation of *Pencillium expansum*, a pathogenic mold .[23]

The fungus contains illudins and cyclopropylglutamine compound called coprine.[24] Active metabolite of coprine, 1-aminocyclopropanol, blocks the action of an enzyme, acetaldehyde, which breaks down acctaldchyde in the body.[25] Acetaldehyde is an intermediate metabolite of ethanol and is responsible for most symptoms of a hangover; its effect on autonomic β receptors is responsible for the vasomotor symptoms.[24]

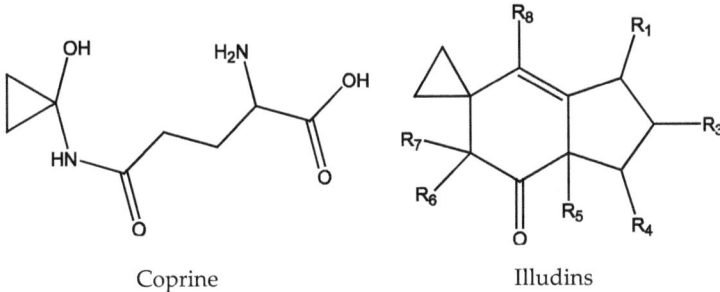

Coprine   Illudins

Coprinus comatus

Common name: Shaggy mane, Shaggy ink cap, Lawyer's wig, Shaggy parasol

Coprinus comatus, is a common fungus often seen growing on lawns, along gravel roads and waste areas. A study of flavor compounds present in C. comatus revealed a variety of compounds in the water extract from the fruit body, including: 3-octanone, 3-octanol, 1-octen-3-ol, 1-octanol, 2-methyl-2-penten-4-olide, 1-dodecanol, caprylic acid, 5-GMP, glutamic acid, n-butyric acid and isobutyric acids.[26,27]

C. comatus possess number of medicinal properties viz. anti-tumor activity[28], hypoglycemic effects[29], antinematode activity[30], antioxidant activity[31]and antimicrobial activity[32].

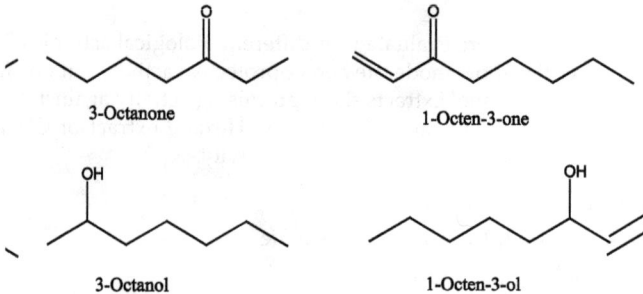

3-Octanone

1-Octen-3-one

3-Octanol

1-Octen-3-ol

Ganoderma lucidum

Common name: Lingzhi mushroom, Reishi mushroom

Ganoderma lucidum is the most popular medicinal mushroom of the world and has been discussed in detail in the proceedings of the symposium. It is reported to posses very significant medicinal properties viz., anti-cancer, anti-HIV, immunomodulatory, hypoglycaemic effects, antibiotic activity, hepatoprotective, anti-atherosclerotic, sleep promoting, anti-aging, and anti-angiogenic. [33-36]

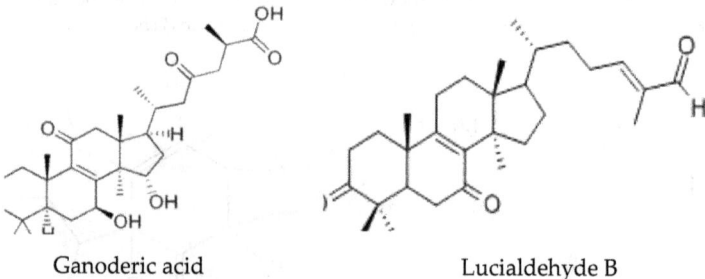

Ganoderic acid

Lucialdehyde B

G. lucidum posses a wide variety of active compon ents, including alkaloids, proteins, amino acids, polysaccharides (including β-D-glucans), ergosterol, ganoderic acids and other sterols, triterpenes, neucleotides (including adenosine), volatile oils, minerals, vitamins and lipids. Some of the active molecules from G.

lucidum include 3α,22β-diacetoxy-7α-hydroxyl-5α-lanost-8,24E-dien-26-oic acid, lucialdehyde B, 8α,9α-epoxy-3,7,11,15,23-pentaoxo-5α-lanosta-26-oic acid and ganoderic acid.[37-41]

Inonotus hispidus

Common name: Shaggy brac ket, Pleated inkcap

Local name: Neeji haddur

Inonotus hispidus (basidiomycetes) is a parasitic fungus preferably living on deciduous trees such as fraxinus, quercus, sorbus and malus (Figure 7)[42]. Previous chemical investigations indicated that *I. hispidus* produced a large variety of yellow-brown polyphenol pigments with a styrylpyrone skeleton, which had been reported to exhibit antimicrobial, antioxidant, antiviral and anti-inflammatory activities.[43-45] Bioactive molecules from *I. hispidus* include inonotusin A and B, (*E*)-4-(3,4-dihydroxyphenyl) but-3-en-2-one, hispidin and 3,4-dihydroxy benzaldehyde.[46]

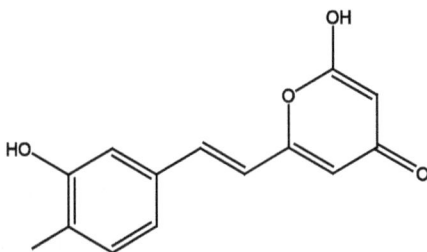

Hispidin

Lactarius controversus

Lactarius is a large genus of mycorrhizal mushrooms that exude a "latex," or milk, when injured and are collectively known as milk-caps (Figure 8). Sesquiterpenes, sterols and other secondary metabolites have been found in Lactarius species.[47, 48]

Various extracts of *Lactarius controversus* showed very promising hepatoprotective activity against paracetamol induced hepatotoxicity on Albino Wistar rats.

Morchella esculenta

Common names: Common morel, Yellow morel, True morel,
Morel mushroom, Sponge morel

Local Name: Kanigech

Morchella esculenta belongs to *Morchellaceae* family of the Ascomycota. It is one of the the edible mushrooms and is most popular of the morels. *M. esculenta* has several medicinal properties, including anti-tumor effects, immunoregulatory properties,[49] fatigue resistance, and antiviral effects,[50-52] antioxidant properties.[53] Aqueous-ethanolic extract of moral mushroom mycelium esculenta, protects cisplatin and gentamicin induced nephrotoxicity in mice.[49]Morel mushroom mycelium extract provides additive or synergistic effect in the prevention and treatment of cancer.[54] About hundred quintals of *M. esculenta* are being collected every year from the forests of Kashmir.

M. esculenta contains wide range of active compounds viz. 5-Dihydroergosterol, Ergosterol peroxide, Ergosterol, Cerevisterol, Trilinolein, *cis,cis*-1,4-pentadiene, γ-glutamyl- transpeptidase enzymes, Methyl myristate, 1-Linoleoylglycerol, Ceramide, Rocaglamide, *cis*-3- amino-l-proline.[55,56]

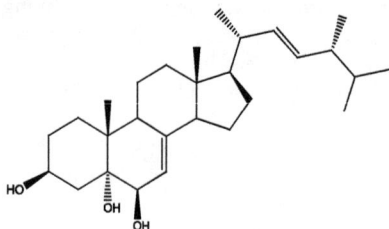

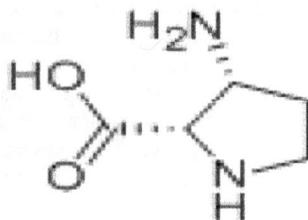

Cerevisterol   *cis*-3- amino-l-proline

Phellinus pomaceus

Phellinus is a genus of fungi, fruiting bodies of which are found growing on wood and are resupinate, sessile and perennial (Figure 9). The name Phellinus means Cork. Phellinus produces the natural phenol Hispidin [57, 58]. Various extracts of *Phellinus pomaceus* showed good hepatoprotective activity against paracetamol induced hepatotoxicity on AlbinoWistar rats.

Russula brevipes

Common name: Short-stemmed Russula

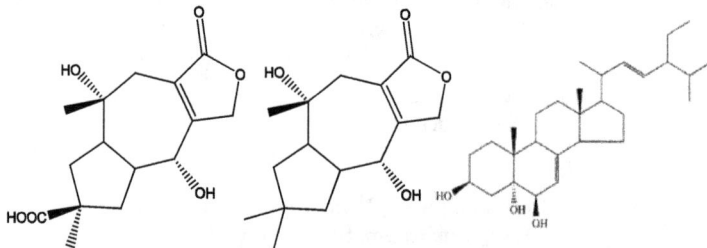

Russulactarorufin   Lactarorufin-A   24-ethyl-cholesta-7,22E-diene-3β,5α,6β-triol

The Short-stalked White Russula (*Russula brevipes*) is a large mushroom in the Russula Family (Russulaceae) (Figure 10). This fungus is edible but is best when parasitized by the ascomycete (a fungus like yeast) *Hypomyces lactifluorum*. It then becomes the highly desirable Lobster Mushroom. Phytochemical investigation in our sister lab.at Jammu led to the isolation of sesquiterpene lactones Russulactarorufin and lactarorufin-A along with 24-ethyl-cholesta-7,22E-diene-3β,5α,6β-triol.[59] Various extracts did not show any promising antibacterial and antifungal activity. Water extract and alcohol:water extract showed activity against HCT-158 (neroblastoma) and HF-295 (colon cancer) cell lines (Table 6).

Trametes versicolor

Common name: Coriolus versicolor, Polyporus versicolor, Polypore mushroom, Turkey tail

Trametes versicolor is a common polypore mushroom found throughout the world (Figure 11). *T. versicolor* is recognized as a medicinal mushroom in Chinese medicine under the name *yun zhi*. Different compounds identified in different extracts of *T. versicolor* include hexadecanoic acid, 5-hydroxy-2-pentanone, lactic acid, furfural, g-butyrolactone, 2-methoxy-4-vinylphenol, 2, 6-dimetoxy-4-vinylphenol and benzaldehyde.[60] In China and Japan, a chemical compound extracted from *T. versicolor*, polysaccharide-K, is used as an Immunoadjuvant therapy for cancer. PSK has shown to be beneficial as an adjuvant in the treatment of gastric, esophageal, colorectal, breast and lung cancers.[61-62]

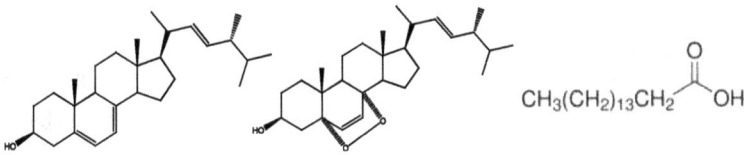

$CH_3(CH_2)_{13}CH_2$ —— OH

Ergosterol   5α,8α-epidioxy-(22E)-ergosta-6,22-dien-3β-ol   Hexadecanoic acid

Future Strategies

Medicinal mushroom industry is expanding fast in view of the use of the MMS as in Immunoadjuvant combination therapies. Future strategies should involve 1) Cultivation of medicinal mushrooms involving conventional and biotechnological interventions;2) Generation of passport data for proper identification, survey and other folklore claims; 3) R & D aspects need to be intensified for development of efficient and modified extraction technologies; 4) Isolation and identification of low molecular weight polysaccharides and their translation for better activity profile; 5) Standardization of medicinal mushroom extracts based on marker compounds; 6)IPR in case of mushrooms to be strengthened.

References

1. Chang, Y. S., and Mills, A. K. *Re-examination of Psilocybe subaeruginosa and related species with comparative morphology, isozymes and mating compatibility studies. Mycol Res, 1992,* **96,**429–441.

2. Wasser, S. P., Medicinal mushrooms as a source of antitumor and immunomodulating polysaccharides. *Appl Microbiol Biotechnol,* 2002, **60,** 258–74.

3. Boa, E. Wild edible fungi: A global overview of their use and importance to people. *Non-Wood Forest Products Series,* 2004, No. 17. FAO, Rome.

4.   Solomon P. W., Medicinal Mushroom Science: History, Current Status, Future
     Trends, and Unsolved Problems. *Int J Med Mushr*, 2010, **12**, 1-16.

5.   Mizuno, T. Development of antitumor polysaccharides from mushroom fungi.
     Food and Food Ingredient Japanese Journal, 1996, **167**, 69–87.

6.   Zhang, M., Cui S.W., Cheung, P.C.K. and Wang, Q. Antitumor polysaccharides
     from mushrooms: a review on their isolation process, structural characteristics
     and antitumor activity. *Trend Food Sci Technol*, 2007, **18**, 4–19.

7.   Nidhi, A., and Chowdhry, P. N., Anti-tumour potential of active compounds
     (polysaccharides) of wild mushrooms from rajouri dist. of Jammu & Kashmir,
     India. *IJPSR.*, 2013, **4**, 11-15.

8.   Mizuno, T. The extraction and development of antitumoractive polysaccharides
     from medicinal mushrooms in Japan. *Int J Med Mushrooms*, 1999, **1**, 9–29.

9.   Kui, Z. and Qiang, W. Optimization of ultrasonic extraction of polysaccharides
     from dried longan pulp using response surface methodology. *Carbohydrate
     Polymers*, 2010, 19–25.

10. Wei, C., Wei, P., Wang, H.S. and Zhang, Q. H., Optimization of ultrasonic-
     assisted extraction of water-soluble polysaccharides from *Boletus edulis* mycelia
     using response surface methodology. *Carbohydrate Polymers*, 2012, **87**, 614–619.

11. Park, H.G., Youn, Y. S., Seung,O.C. and Won,M.P., New method development
     for nanoparticle Extraction of water-soluble β-(1◦3)-D-Glucan from edible
     mushrooms, *Sparassis Crispa* and *Phellinus Linteus*, *J. Agric. Food Chem.* 2009,
     57, 2147–2154.

12. *Chanchal, K., Nandan, Ramsankar, S., Sunil, K., Bhanja, S. R. and Syed, S. I., Isolation
     and characterization of polysaccharides of a hybrid mushroom (backcross mating
     between PfloVv12 and Volvariella volvacea). Carbohydrate Res.* 2011,*346*, 2451-2456.

13. Pai-Feng,K., Shwu-Huey,W.,Wei-Ting,H.,Yu-Han, L., Chun-Mao, L. and
     Wen,B.Y., Structural Characterization and Antioxidative Activity of Low-
     Molecular-Weights Beta-1,3-Glucan from the Residue of Extracted *Ganoderma
     lucidum* Fruiting Bodies. *J Biomed Biotechnol.* 2012, **673764**, 8.

14. *Colombo, M.L., and Bosisio, E., Pharmacological activities of Chelidonium majus
     (Papaveraceae). Pharmacol. Res.,* 1996, *33*, 127 – 134.

15. *Chan, Y., Chang, T., Chan, C.H., Yeh, Y.C., Chen, C.W., Shieh, B. and Li, C.,Immuno-
     modulatory effects of Agaricus blazei Murril inb Balb/cByJ mice. J. Microbiol. Imuunol.
     Infect.* 2007,*40*, 201–208.

16. Rames, J. N., Kumar,S., Ishaq, M., and Kumar, H., A review on hepatoprotective
     activity of medicinal plants. *IJPSR*, 2011, **2**,501-515.

17. Tetsuro, I., Nobuaki, U., and Yuko, M., *Antitumor activity of aqueous extracts of
     edible mushrooms. Cancer Res*, 1969, **29**, 734-735.

18. Tomasi, S., Lohezic, F., Sauleau, P., Be´zivin, C., and Boustie, Cytotoxic activity
     of methanol extracts from Basidiomycete mushrooms on murine cancer cell
     lines. *J. Pharmazie.* 2004, **59**, 290–293.

19. Sasek, V. and Musflek, V., Cultivation and antibiotic activity of mycorrhizal basidiomycetes. *Folia Microbiol.* 1967, **6**, 515-523.

20. Sophie, B., Thierry, T., Sylvie, R., and Jean,B., Monoterpenes in the Aromas of Fresh Wild Mushrooms (Basidiomycetes).*J. Agric. Food Chem.*, 1997, **45**, 831–836.

21. *Ohtsuka, S., Ueno, S., Yoshikumi, C., Hirose, F., Ohmura, Y., Wada, T., Fujii, T. and Takahashi, E., Polysaccharides having an anticarcinogenic effect and a method of producing them from species of Basidiomycetes. UK Patent 1331513, 26 September 1973.*

22. Lee I.K., Jeong, C.Y. and Cho, S., Illudins C2 and C3, new illudin C derivatives from *Coprinus atramentarius* ASI20013. *J Antibiot.*, 1996, **49** 821-2.

23. Benjamin, D.R., *Mushrooms: poisons and panaceas*, 1995, pp. 286

24. Marchner, H., and Tottmar, O., A comparative study on the effects of disulfiram, cyanamde, and 1-aminocyclopropanol on the acetaldehyde metabolism in rats. *Acta Pharmacologica Toxicologica*, 1978, **43**, 219–32.

25. Duncan, C.J.G., Pugh, N., Pasco, D.S. and Ross, S.A., Isolation of a galactomannan that enhances macrophage activation from the edible fungus *Morchella esculenta. J Agric Food Chem,*,2002, **50**, 5683–85.

26. *Dijkstra, F.Y. and Wiken, T.O., Studies on mushroom flavours. Flavour compounds in Coprinus comatus. Z Lebensm Unters Forsch, 1976, 160, 263-9.*

27. *Gu, Y. H. and Leonard, J., In vitro effects on proliferation, apoptosis and colony inhibition in ER-dependent and ER-independent human breast cancer cells by selected mushroom species. Oncol Rep., 2006, 15, 417-23.*

28. *Han, C., Yuan, J., Wang, Y. and Li, L., Hypoglycemic activity of fermented mushroom of Coprinus comatus rich in vanadium. J Trace Elem Med Biol., 2006, 20,191-6.*

29. *Li, Y. and Xiang, H., Nematicidal activity of Coprinus comatus. Acta Phytopathologica Sinica, 2005, 35, 456-8.*

30. *Badalyan, C.M., Gasparyan, A.V. and Garibyan, N.G., Investigation of the antioxidant activity of some basidial macromycetes. Mikol Fitopatol., 2003, 37,63-8.*

31. *Ershova, E.Y., Efremenkova, O.V., Zenkova, V.A., Tolstykh, I.V. and Dudnik, Y.V., The revealing of antimicrobial activity of strains of the genus Coprinus. Mikol Fitopatol., 2001, 35,32-7.*

32. Yaoita, Y., Endo, M., Tani, Y., Machida, K., Amemiya, K., Furumura, K. and Kikuchi, M., Studies on the constituents of mushrooms, part VI-Sterol constituents from seven mushrooms. *Chem Pharm Bull.*, 1999, **47**, 847-51.

33. Wang, J., Wang, Y., Liu, X., Yuan, Y. and Yue, T., Free radical scavenging and immunomodulatory activities of *Ganoderma lucidum* polysaccharides derivatives. *Carbohydr Polym.*, 2013, **91**, 33-8.

34. Li, Y., Zhu, Z., Yao, W. and Chen, R., Study progress on triterpenoids from *Ganoderma lucidum. Zhongguo Zhong Yao Za Zhi.*, 2012, **37,**165-71.

35. Sanodiya, B.S., Thakur, G.S., Baghel, R.K., Prasad, G.B. and Bisen, P.S., *Ganoderma lucidum*: a potent pharmacological macrofungus. *Curr Pharm Biotechnol,*. 2009, **10**,717-42.

36. Soniamol, J., Janardhanan K.K., George, V. and Sabulal B., A new epoxidic ganoderic acid and other phytoconstituents from *Ganoderma lucidum*. *Phytochem Lett*, 2011,**4**, 386–388.

37. Vincent, E. C., and Fang, L., Immunomodulation and Anti-Cancer Activity of Polysaccharide-Protein Complexes. *Currt Med Chem*, 2000, **7**, 715-729.

38. Russell, R. and Paterson, M., *Ganoderma* – A therapeutic fungal biofactory. *Phytochem*, 2006, **67**, 1985–2001.

39. Jin, X., Ruiz,B. J., Sze, D.M. and Chan, G.C., *Ganoderma lucidum* (Reishi mushroom) for cancer treatment..*Cochrane Database Syst Rev.*, 2012, 13, 6.

40. Boh, B. *Ganoderma lucidum*: A Potential for Biotechnological Production of anti-cancer and Immunomodulatory drugs. *Recent Pat Anticancer Drug Discov.* 2012.

41. Jonsson, M., Lindquist, N.G., Ploen, L, Ekvarn, S. and Kronevi. T., Testicular lesions of coprine and benzcoprine. *Toxicol.* 1979, **12**,89-100.

42. Ali, N., Jansen, R., Pilgrim, H., Liberra, K. and Lindequist, U., Hispolon, a yellow pigment from *Inonotus hispidus*. *Phytochem*, 1996,**41**, 927—929.

43. Ali, N., Mothana, R., Lesnau, A., Pilgrim, H. and Lindequist, U., Antiviral activity of *Inonotus hispidus*. *Fitoter*, 2003,**74**, 483—485.

44. Ali, N., Ludtke, J., Pilgrim, H. and Lindequist, U., Inhibition of chemiluminescence response of human mononuclear cells and suppression of mitogen-induced proliferation of spleen lymphocytes of mice by hispolon and hispidin. *Pharmazie*, 1996, **51**, 667—670.

45. Wang, Y., Shang, X. Y., Wang S. J., Mo, S. Y., Li, S., Yang, Y. C., Ye, F., Shi J. G. and He, L., Structures, biogenesis, and biological activities of pyrano[4,3-c] isochromen-4-one derivatives from the Fungus *Phellinus igniarius.J. Nat. Prod.*, 2007, **70**, 296—299.

46. Xiao, C., Wu, Q.P., Cai, W., Tan, J.B., Yang, X.B. and Zhang, J.M., Hypoglycemic effects of *Ganoderma lucidum* polysaccharides in type 2 diabetic mice. *Arch Pharm Res.* 2012, **35**,1793-801.

47. *Vidari, G., and Vita-Finzi, P. Sesquiterpenes and other secondary metabolites of genus Lactarius (Basidiomycetes): Chemistry and biological activity. Studies in Natural Products Chemistry, 1995, 17, 153-206.*

48. *Cerri, R., Simone, F. D. and Senatore, F., Sterols from three Lactarius species. Biochemical Systematics and Ecology, 1981, 9, 247-248.*

49. Nitha, B. and Janardhanan, K.K., Aqueous-ethanolic extract of morel mushroom mycelium *Morchella esculenta*, protects cisplatin and gentamicin induced nephrotoxicity in mice. *Food Chem Toxicol*, 2008, **46** , 3193–99.

50. Rotzoll, N., Dunkel, A. and Hofmann. T., Activity-guided identification of (S)-malic acid 1-O-D-glucopyranoside (morelid) and gamma-aminobutyric acid as contributors to umami taste and mouth-drying oral sensation of morel mushrooms (*Morchella deliciosa* Fr.). *J Agric Food Chem*, 2005, **53**, 4149–56.

51. Wasser, S.P., Medicinal mushrooms as a source of antitumor and immunomodulating polysaccharides. *Appl Microbiol Biotechnol,* 2002, **60,**258–74.

52. Mau, J.L., Chang, C.N., Hunag, S.J. and Chen, C.C., Antioxidant properties of methanolic extracts from *Grifola frondosa, Morchella esculenta* and *Termitomyces albuminosus* mycelia. *.Food Chem,* 2004, **87** , 111–18.

53. Jeong, A.K., Edward, L., David, T., Esperanza, J. and Carcache, D., Antioxidant and NF-kB inhibitory constituents isolated from *Morchella esculenta. Natl Product Res,* 2011,**25**, 1412–1417.

54. Nitha, B., Meera, C. R., and Janardhanan, K. K., Anti-inflammatory and antitumour activities of cultured mycelium of morel mushroom, *Morchella esculenta. Curr Sci,* 2007, **92,** 235-239.

55. Duncan, C., Pugh, J. G., Pasco, D. S. and Ross, S. A., Isolation of a galactomannan that enhances macrophage *activation* from the *edible fungus Morchella esculenta.* *J Agricl Food Chem,* 2002, **50,** 5683–5.

56. Tyler, V. E. and Groger, J. D., Amanita alkaloids. II. Amanita citrina and Amanita porphyria *Planta Med,* 1964, **12,** 397-402.

57. Ellis, J. B., Ellis, and Martin, B. *Fungi without gills (hymenomycetes and gasteromycetes): an identification handbook.* London: Chapman and Hall. 1990, ISBN 0-412-36970-2.

58. Lee I.K., and Yu B.S., Highly oxygenated and unsaturated metabolites providing a diversity of hispidin class antioxidants in the medicinal mushrooms *Inonotus* and *Phellinus. Bioorganic Med Chem,* 2007, **15**, 3309-3314.

59. Suri, O.P., Shah, R., Satti, N.K. and Suri, K.A., Russulactarorufin, a lactarane skeleton sesquiterpene from *Russula brevipes. Phytochem,* 1997, **45**, 1453–55.

60. Masumi, K., Masahiro, H., Katsumi, U., Kazuo, K., Yuzuru, O. and Takayuki, S., Antioxidant/anti-inflammatory activities and chemical composition of extracts from the mushroom *Trametes versicolor. Int J Nutr Food Sci,* 2013, **2,** 85-91.

61. Fisher, M. Y., Anticancer effects and mechanisms of polysaccharide-K (PSK): implications of cancer immunotherapy. *Anticancer Res,* 2004, **22**, 1737–1754.

62. Dijkstra, F.Y., Studies on mushroom flavours, some flavour compounds in fresh, canned and dried edible mushrooms. *Z Lebensm Unters Forsch,* 1976, **160,** 401-5.

5

Chinese caterpillar mushroom *Cordyceps sinensis*: Historical, collection, culture and medicinal use for cure of human diseases

B.L. Dhar *, N. Srivastava, Himanshu, Jitindra, Sonika and Priyenka
Mushroom Research Development and Training Centre, DK Floriculture, Usha Fasrm, Bhijwasan, Delhi-110061. India. *e-mail: beharilaldhar@gmail.com

Cordyceps sinensis, an entomophagus insect parasitic fungus in the alpine region has been highly valued in traditional medicinal system of China, Nepal, Tibet and India. The first report of positive effect of this mushroom was discovered about 1500 years ago in Tibet by herdsmen. The first Chinese Emperor used this herbal medicinal mushroom as a tonic for longevity of life. *C. sinensis* is an important anti-aging medicine, as it inhibits the formation of active monoamine oxidase, an enzyme responsible for aging in man and is also used for lung protection, various reproductive disorders. There is a large and fast growing market worldwide for *C. sinensis* as a medicine. Attempts at its cultivation under controlled environment artificially have not succeeded so far. Nevertheless, two methods mainly used in the cultural propagation of this mushroom fungus, liquid/fermentation method and solid-substrate method. The liquid culture multiplication is more popular in China and solid substrate mycelial multiplication is popular in USA and Japan. The medicine preparation procedure from *C. sinensis* needs to be standardized, so that similar products from all manufacturers yield similar results on analysis.

Introduction

Mushrooms from time immemorial have been linked to mystery and man has always been suspicious about its usage. But with more and more information coming together about its biology, nature and food/medicinal use, it is becoming a potent tool in the hands of the researchers for cure of terminal diseases like cancers/ other ailments. Deng Chong Xia Cao, winter worm summer grass, the Chinese caterpillar mushroom *Cordyceps sinensis* is gaining importance as a useful source of medicine for critical human disease control in the world today (Figure 1). Locally this mushroom is known as 'Yarsha Gumba', Yarsha Gamboo or 'Kira Ghas' since it parasitizes the Lepidopteran insect larvae of the *Hepialus armoricanus*, family Hepialidae under the soil[1]. The first ever report of this fungus dates back to 18th century when Torrubia, a Franciscian friar in Cuba described it as the tree growing out of the bellies of wasps. That is the reason why genus *Cordyceps* is sometimes referred to as the *Torrubia* in honour of its inventor[2]. The combination of an insect larvae or rather the mummy of a larvae converted into fungus mycelium along with the fungus stalk is known as 'Chinese plant worm' in the ancient literature, and now named as *C. sinensis*. The first report of positive effect of this mushroom was discovered about 1500 years ago in Tibet by herdsmen, who observed that their livestock became energetic after consuming a special mushroom growing locally in grasslands. Almost a thousand years later, the physicians of the Emperor of Ming Dynasty came to know about this wonderful mushroom and used this knowledge combined with their own wisdom to develop new and powerful medicines. Earliest records of *Cordyceps* as medicine is as old as the Qing dynasty in China as mentioned in Ben-Cao-Cong-Kin, a new compilation of Materia Medica written by Wu-Yiluo in 1757[3]. *C. sinensis*, an entomophagus insect parasitic fungus in the alpine region has been highly valued in traditional medicinal system of China, Nepal, Tibet and India. This mushroom is valued 4 times its weight in silver, and is used for lung protection, various reproductive disorders and also balancing "Qi" the fundamental energy of life [4]. Being a hormone stimulator, *C. sinensis* is an important anti-aging medicine, as it inhibits the formation of active monoamine oxidase, an enzyme responsible for aging in man. The first Chinese Emperor used this herbal medicinal mushroom as a tonic for longevity of life. The legendary Chinese beauty 'Yang Kue Fei" (719-756 AD) reported to be a regular user of this medicinal herb, crediting it as fountain of youth. The Chinese athletes consumed this fungus regularly and held many world records[5]. It was chiefly used to strengthen the body after exhaustion or long term illness. Later it was used for control of disorders like backache, impotency and other related problems like opium addiction, etc.[6,7].

Fig. 1: *Cordiceps sinensis (Smith et al. 2001)*

The medicinal use of *Cordyceps sinensis* by Tibetans has been documented for over 50 years. The fruiting bodies and mycelium culture of *C. sinensis* have been and are still used for strengthening the immune system[8] and for many types of ailments of heart, kidney, liver, circulatory system and for treating TB, asthma, back pain, reproductive disorders, cancers, etc.[9-10] apart from the treatment of Hepatitis B[11]. It is a traditional medicine with a long history of use and is now ready to be used on an ever increasing scale in the treatment of more and more diseases. There is a large and fast growing market worldwide for *C. sinensis* as a medicine.

Habitat

C. sinensis is distributed in nature in various regions of the world and is reported from China, Japan, Germany, USA, Mexico, Canada, Denmark and Italy. Native occurrence of this mushroom is confined to the high Himalayan Mountains in Tibet plateau (China), Nepal and India, at an altitude ranging from 3000 to 5000m and in cold and arid environment. In India, it is collected from upper hilly regions of district Pithoragarh in Uttrakhand State in northern hilly regions adjoining China, at an altitude of 3200m from the snow meadows of Brahamkot, Ultapara, Ghawardhappa, Chipalakot, Najari (Darachula), Chatri Bugyal and Chipla, Kedar at an altitude of 4000m above sea level [12]. It has also been reported to be collected from Nagin Dhura, Ralan Bugyal at the base of Panchachuli hills, Laspa, Tola top, Darti, Mapa top, Burfu top, Milan top in Johar hills of Kumaon from Uttranchal State[13]. In China this mushroom is reported growing naturally in the cold higher mountains at 3000m – 5000m above sea level in Sichuan, Qingtar, Xizang, Yunnan, around the Himalayas [14]. The distribution of this mushroom is limited to areas with an average annual rainfall of about 350mm – 400mm in general. *Cordyceps* has not been reported from areas where annual precipitation is below 300 mm like Chang-Tang and other arid areas of the north western plateau in China.

C. sinensis is named differently in different regions/ countries, viz; "vegetable wasps and plant worms" in western countries, 'Yartsa Gunbu' in Tibet and 'Gadavira' in Nepal[15]. In Dolpa area of Nepal, it is known by many names such as Yarsagumba, Jara(root), Kira(insect), Jeevan Buti chyan(Life tonic mushroom) and Chyan kira-Mushroom insect[16]. In Pithoragarh, India it is also known by the name of 'Ghas ka kira' or 'Kira Jhar' or 'kira ghas'. It is known as 'Dong Chung Ha Cao' or 'winter worm summer grass' in Korea, 'Totsu kasu' or 'Tochukasu' in Japan and 'Dong Chong Xia Cao' in China[17].

Cordyceps sinensis mushroom- attempts at its cultivation

The fruit body or ascocarp of *C. sinensis* mushroom originates at the base on an insect larvae host, usually the larvae of the 'Himalayan Bat Moth' *Hepialus armoricanus*, and ends at the club like cap, including the stipe and stroma[17]. The fruiting body of caterpillar fungus consists of head parts and parts that look like sacks. The fruit body is dark brown to black, and the 'root' of the organisms (larvae body) pervaded by the mushroom mycelium and appears yellowish to brown in colour[17]. In Tibetan medical literature it is described as one with slender, short and bamboo-shaped root which has smoke or ochre colour with light yellowish or skin colour exterior and white fleshy interior. The root has worm-like head, body and legs with numerous thin and fine transverse wrinkles. There are about 8 pairs of legs on the body of the root and out of these four middle pairs are more prominent.

Its lower part is thin while the upper part is slightly thicker[18]. The immature larvae which forms the host upon which the fungus *C.sinensis* grows, usually lives about 6" below ground[17]. From the collar of this solitary root grows a dark brown grassy stalk which is thickened at the middle with slightly pointed tip and slender base[18]. The length of the mushroom also differs greatly, from a few mm to 10 mm[19].

The fungus *C. sinensis* parasitizes the larvae of a small moth, *H. armoricanus*. The mycelium of this fungus grows in soil and colonizes the buried larvae/ caterpillar of this moth producing insect. The caterpillar becomes mummified by the growth of the mycelium, and later the fungus forms a fruiting body, which emerges from the head of the larvae [12,17]. *C sinensis* parasitizes a range of grass root boring caterpillars, which would hatch as whitish ghost moths, when not infected by *C. sinensis*. The normal reproductive cycle for *Hepialus armoricanus* takes upto 5 *years*, and most of the cycle is lived as a caterpillar, and the host for *C sinensis* most commonly reported is *H. armoricanus*. Other host larvae have been identified [20,21]. About 40 species of *Hepialus* moth are recognized in the 'Tibetan Plateau' region and according to Chen *et al*, 30 species can be infected by *C. sinensis*. Three hosting types of *Hepialus* are noted:

    i)   Xi Zang – king Cordyceps, grows in Tibet, Qinglai and Sichuan. It is the largest sized mushroom sold in market. It has a distinct caterpillar body with a thick tail. (Tibetan *Cordyceps*/ preferred *Cordyceps*).

    ii)   Qing Hai-aromatic of the three types.

    iii) Si Chuan –smaller, dark brownish body, with ordinary aroma.

The above three aspects of fruit body size, aroma and colour are considered as a key quality criteria[22].

Attempts at its cultivation under controlled environment artificially have not succeeded so far. Caterpillars of other insects/ silk worm were tried as hosts but without any success .The main sources of the *C. sinensis* medicine are the fruit bodies collected from wild and mycelial culture products. Commercially viable methods for its mycelial multiplication were finally developed in late 1970s. By mid 1980s, the majority of *Cordyceps* medicine available in the world market was from mycelial culture, grown either in liquid or solid media/ or what is popularly known in China as fermentation method.

Cordyceps sinensis mycelial culture propagation

There are two methods mainly used in the cultural propagation of this mushroom fungus, liquid/fermentation method and solid-substrate method. Liquid/fermentation method is mainly used in China wherein the fungus is grown in sterilized nutritionally rich liquid medium inside fermenters, filtered and the mycelial matter is harvested and used for medicine preparation. But most of the exo-enzymes (extra cellular compounds) that the fungus produces in the medium are lost in the filtrate solution.

Second method of mycelial cultivation is the solid substrate method, where the mushroom fungus is multiplied on sterilized solid substrates like wheat/ rice grains. This method is followed by most growers in Japan and USA. The method used is similar to spawn preparation method used in edible mushroom cultivation. The solid substrate after boiling is sterilized in glass jars/ pp bags, inoculated with pure culture fungus and then incubated. The entire material, fungus + solid substrate is dried and powered, and used for preparation of the *Cordyceps* medicine.

Cordyceps sinensis medicinal properties

Medicinal properties of *C.sinensis* are attributed to Cordycepin, Cordycepic acid, triterpenoids and other active compounds [23]. *C. sinensis* is well known as Chinese Medicinal 'Herb' used for its invigorating and immunological effects on the human body[24]. In literature the word *Cordyceps* generally means *C. sinensis*, is a popular Chinese tonifying herb, and is being revered for being both ' -nourishing' and Yang invigorating in Chinese. For medication, the fruiting body (mushroom) and the worm (caterpillar) are used together, and worm has chemical composition similar to the fruiting body. Internationally the health efficacies of *Cordyceps* were observed and tested against asthma, allergic rhinitis, poor renal function, renal injuries caused by chemicals, chronic bronchitis/ coughing, poor resistance of respiratory tract, regulation of blood pressure (high & low), anti aging, weakness, the declining sex drive, lowering raised serum lipid levels, strengthening the body's immunity/ improving the poor function of lungs and kidneys, and in irregular menstruatio ns [3,25-27].

Cordyceps – the Chinese herbal medicine and clinical trials

Some clinical trials reported are as under:

Active component	Medicinal application
Cordycepin	- Antitumor – inhibition of RNA/DNA synthesis[28] - Suppression of viral replication[29] - Anti HIV[30] - Anti malaria[31-33]
Adenosine	- Anti inflammatory[34] - Blood flow control[35] - Prevention of cardiac arrhythmias
Amino acids, Zinc, vitamins and trace elements	- Combats sexual dysfunction[36]
Polysaccharides	- Lipid peroxidation inhibition[37,38] - Prevention of hemolysis and inhibition of tumors[39] - Anti oxidant activity
Ergosterol	- Antitumor and immunomodulatory effect[33]
Cordyglucans	- Antitumor activity[40]

Role of *Cordyceps* mushroom in development of economy in the region of its occurrence

In last 15-20 years, *C. sinensis* mushroom collection from wild habitats has gained momentum and this mushroom is being traded very extensively in Dharchula and Mansyari areas in district Pathoragarh of Uttrakhand State, India[41]. Trades have been visiting these areas from Nepal or Tibet to buy *Cordyceps* locally in the villages of these districts. There is no organized trade for collection/marketing of this mushroom in the areas of its occurrence. For each mushroom the villagers are offered Rs.30.00 or more and approximately 3000-3500 mushrooms make one kilogram (dry). The estimated trade volume from above mentioned areas is around 1000 kg/annum when one kg fetches from Rs.1-1.5 lakhs (1 lakhs = Rs100,000). The natural resource has some impact on the rural economy in these areas. During the month of May and June, the young villagers camp in these areas for collection of mushrooms. All young boys spend 2 months of summer in these mountains to collect this expensive mushroom from the wild. Families of 4-5 persons collect about half to one kg mushrooms in one season which helps in subsisting their family incomes to a great extent[41].Traditionally this fungus is traded in China for its weight in silver and gold, and the market price is astronomically high. It is believed that in international market this mushroom may fetch a price of US $20000-40000 per kg of mushrooms[42]. The current prices of mushroom per kg ranges from Indian Rs 68000-80000 in Tibet, Rs 80000-90000 in Nepal, Rs 1.25- 1.3 lakhs in India[12].

There is a long list of active ingredients/ constituents of medicinal significance in the Chinese caterpillar medicinal mushroom *C. sinensis* and these are briefly as under:

Cordyceps sinensis mushroom world trade, mycelium manufacture and marketing

There is a large and fast growing market worldwide for this medicinal mushroom, as a medicine and as a health supplement. The increased demand and high prices of this mushroom have stimulated the workers for the development of its cultivation methods. Till date there has been no organized result oriented attempt at its cultivation under controlled conditions. Due to these reasons more and more wild mushrooms are being harvested from nature every year. The high prices of wild mushroom *C.sinensis* have led unscrupulous harvesters to resort to unfair means of inserting twigs or even lead wires into the stomata for increasing the weight of the fruit body for more money. Laboratory culture of *C.sinensis* is typically by growth of fungus in pure culture either on liquid or solid substrate media for medicine preparation. The mycelium is grown in fermentors in liquid medium/ solid substrate and mycelium harvested for medicine preparation. The liquid culture multiplication is more popular in China and solid substrate mycelial multiplication is popular in USA and Japan. The fungus after bringing into pure culture is multiplied like spawn is prepared for edible mushroom cultivation.

The world marketing of *C. sinensis* medicine is done by various companies from Japan, China, USA, UK, Australia and other countries. The companies sell this medicine purely as mycelial product or a mixture of fruit body and mycelium under various trade names. The pharmacological work needs to be strengthened in this area as products from various companies of this medicine do not show uniformity in content.

Future thrust areas in development of this mushroom and its products for human welfare

This mushroom requires to be declared as a protected commodity and the random harvest of this mushroom prevents its preseverence in its natural habitats.

Serious efforts are required to be made jointly by mycologists and entomologists to find alternative insect hosts for development of this mushroom under controlled conditions.This mushroom shows promise for the treatment of so many diseases, but we do not have the back up data on the active ingredients involved either alone or in combination. It is the whole mushroom which is a very potent in disease control as compared to its mycelial product. More investigations in these areas will be very revealing and beneficial for the human beings for disease control. The mycelial multiplication methods, though simple require be standardizing, as in spawn making for edible mushrooms, and popularizing for production of this medicine at competitive/ affordable rates. The medicine preparation procedure from *C.sinensis* needs to be standardized, so that similar products from all manufacturers yield similar results on analysis. This is very important from quality point of view and the international trade.

References

1. Huang, K.C., Tonics and supporting herbs. In: *The Pharmacology of Chinese Herbs.* (Huang, KC-Ed): (CRC Press; Boca Raton. London, NY, Washington DC), 1999, 263-264.

2. Christensen, C.M., Fungus predators and parasites. In: *Moulds, Mushrooms and Mycotoxins,.* (Univ. of Minn Press, Minneapolis,USA), 1975, pp 164.

3. Zhu, J.S., Halpern, G.M. and Jone, K., The scientific rediscovery of an ancient Chinese herbal medicine *Cordyceps sinensis*. *J. Altern. Complement Med.* 1998, **4**, 298-303.

4. Jones, An ancient Chinese secret promotes longevity and endurance. *Healthy and Natural J.* 1997,**3**, 90-93.

5. Steinkraus, D.C., Whitefield, J.B., Chinese caterpillar fungus and world record runners. *Am.Entomol.* 1994, 40, 235-239.

6. Zhu, J.S., Halpern, G.M. and Jones, K., The Scientific Rediscovery of an Ancient Chinese Herbal Medicine : *Cordyceps sinensis*, Part I, *J. Alt. Compl. Med.* 1998, **4**, 289-303.

7. Zhu, J.S., Halperen, G.M. and Jones, K., The Scientific Rediscovery of an Ancient Chinese Herbal Medicine: *Cordyceps sinensis* Part-II, *J. Alt. Compl. Med.*, 1998b,**4**, 429-457.

8. Hong, Z, and Li, Y., Immuno pharmacological function of *Cordyceps*. *Chinese J. Int. Trade Western Med.* 1990,**10**, 570-571.

9. Hobbs, Christopher, L.A.C. Medicinal Mushrooms:An Exploration of Tradition, Healing and Culture. *Botanica Press*, Santa Cruz, CA, 1995, 81-86.

10. Tusboi, H., Tusnoo, A., Kimjo, N., Nian, Lai, H. and Miyamoto, H., Effects of the mycelial extract of cultured *Cordyceps sinensis* in *in vivo* hepatic energy metabolism in the mouse. *Jpn. J. Pharmacol*, 1996,**70**, 85-88.

11. Manabe, N., Sugimoto, M., Azuma, Y., Taketomo, N., Yamashita, A., Tsuboi, H., Tsunoo, A., Kinjo, N., Nian-Lai, H., and Miyamoto, H., Effects of the mycelial extract of cultured *Cordyceps sinensis* on in vivo hepatic energy metabolism in the mouse. Jpn J Pharmacology, 1996, **70**, 85 – 88.

12. Zhon, L., Short term curative effect of cultured *C.sinensis* (Berk) Sacc mycelia in chronic hepatitis-B. *Chang Kuo Chung Yao, Tsa Chih,* 1990,**19,** 53-55.

13. Sharma, Subrat, Trade of *Cordyceps sinensis* from high altitudes of the Indian Himalayas conservation and biotechnological priorities. *Curr.Sci.* 2004,**86,**1614-1619.

14. Negi, P.S., Systematics of Mushrooms of Almora and Pithoragarh districts. Ph.D thesis, Deptt. of Botany, Kumaon University, Almora, 2005,162 pp.

15. Gwang-Po Kim., Characteristics of *Cordyceps* species. Hits. www. mushroomworld.com. 2000, 780

16. Hye-Young Lee. The life of caterpillar fungi and its cultivation. hits: www. mushroomworld.com. *Planta Medica.* 1999,383.

17. Devkota, S., Yarsagumba *Cordyceps sinensis* (Berk.) Sacc. Traditional utilization in Dolpa district,Western Nepal. *One Nature,* 2006,**4,**48-52.

18. Holiday, J. and Cleaver, M., On the trail of the Yak: Ancient *Cordyceps* in the modern world (In Press).www.mushroomworld.com/medicine, 2004.

19. Gerbyal, S.S., Aggarwal, K.K., Babu, C.R., Impact of *Cordyceps sinensis* in the rural economy of interior villages of Darachula sub-division of Kumaon Himalayas and its implications in the society. *Ind. J of Traditional Knowledge,* 2004,**3,** 182-186.

20. Hye-Young lee, The life of caterpillar fungi and its cultivation. hits: website: *wwwmushroomworld.com* . 1999. 383.

21. Gao, Z.X., Chen, J., Yu. H., Zhao, Z.H. and Song, H.M., Study on the main host swiftmoth, *Hepialus oblifurcus* Chu et Wang, of the caterpillar fungus *Cordyceps sinensis* (Berk) W Sacc in Kangding. *Acta – Entomologica – sinica,* 1992,**35,** 317-321.

22. Chen,Y.Q., Wang, N., Zhou, H. and Qu, L.H., Differentiation of medicinal *Cordyceps* species by rDNA 18s sequence analysis . *Planta Medica,* 2002,**68** 635-639.

23. Vinning, Grant, and Tobgay, Sonam, *Cordyceps*: a market analysis. *Min. of Agri., Thimphu, Bhutan.* www.mac.gov. bt/mac/reports/reports/*Cordyceps*.doc. 2004.

24. Rana, V.S., Propagation prospects of caterpillar mushroom. *Natural Product Radiance.* 2004, **3,** 167-169.

25. Jiang, Y. and Yao, Y.J., Names related to *Cordyceps sinensis* anamorph. *Mycotaxon*: Oct – Dec. 1984: 245-254.

26. Francia, C., Rapor, S., Courtececuisse, R. and Sironx, Y. Current Research Findings on the effects of selected mushrooms in cardiovascular diseases. *Int. J. Med. Mushr.* 1999,**1,**169-172.

27. Halpern, G.M., *Cordyceps*, Chinese Healing Mushroom. *Avery Publishing Group,* New York, USA, 1999, pp.16.

28. Mizuno, T., Medicinal Effects and Utilization of *Cordyceps* (Fr.) Link (Ascomycetes) and Isaria Fr. (Mitosporis fungi) Chinese caterpillar fungi "Tochukaso". *Int. J. Med. Mushr.* 1999, **1**: 251-262.

29. Penman, S. and Roseback, M., Messenger and Heterogenus nuclear RNA in cells: differential inhibition by *Cordyceps. Proc. Natl. Acad. Sci.*, USA, 1970, **67**, 1878-1885.

30. Trigg, P., Gutteridge, W.E. and Williamson, J., The effect of *Cordyceps* on material parasites. *Trans. R. Soc. Trop. Med. Hyg.* 1971,**65**, 514-520.

31. Mucller,W.E.G.,Weiler,B.E.,Charubala,R.,Pfleiderer,W.,Laserman,L.,Sabol,R.W. ,Suhadolnik, R.J. and Shroeder,H.C., Cordycepin analogues –2'5'oligadenylate inhibit human immunodeficiency.Virus infection inhibition of reverse transcription. *Biochem Wash.* 1991,**30**, 2027 – 2033.

32. Kuo, Y.C., Lin, C.Y., Tsai, W.J., Wu, C.L., Chen, C.P., Shiao, M.S., Growth inhibitors against tumor cells in *Cordyceps sinensis* other than Cordycepin and polysaccharides. *Cancer Invest*, 1994,**12**, 611-615.

33. Miller, R.A., Lab-grown *Cordyceps sinensis* hybrid. A nano-processed Medicinal Mushroom that really delivers. www.pharmacenticatabolismnwbotanicals.org. 2005.

34. Ng,T.B. and Wang, H.X., Pharmacological actions of *Cordyceps*, a prized folk medicine; *J.Pharm Pharmacol.* 2005,**57**,1509-1519.

35. Berne, R.M., The role of adenosine in the regulation of coronary blood flow. *Circ. Res.* 1980, **47**, 807-813.

36. Pelleg, A. and Porter, R.S., The pharmacology of Adensine. *Pharmacother.* 1990,**10**,157-174.

37. Yang, H.D., Ma, Z.L., Suo T.Q., Zhang, X.C. Cai, J.G., Comparative study on the chemical constituents between Xiangbangchangcao (*Cordyceps barnessi) and Cordyceps (C.sinensis), Chin Trad. Herban Drugs,* 1985,**16**,194-195

38. Li, S.P., Zhao, K.J., Ji, Z.N., Song, Z.H., Dong, T.T.X., Lo, C.K., Chenng, K.H., Zhu, S.Q. and Tsim, K.W., A polysaccharide isolated from *Cordyceps simensis* a traditional Chinese medicine, protects PC 12 cells against hydrogen peroxide induced injury. *Life Sci.* 2003.**73**, 2503-2513.

39. Li, S.P., Li, P., Dong, T.J. and Tsim, K.W.K., Antioxidation activity of different types of natural *Cordyceps snmensis* and culture *Cordyceps* mycelia *Phytomed.* 2001,**8**,207-212.

40. Yamaguchi, Y., Kagota, S., Nakamura, K., Shinozuka, K. and Kunitomo, M., Antioxidant activity of the extracts from fruiting bodies of cultured *Cordyceps sinensis. Phytother. Res.* 2000,**14**,647-649.

41. Wu, Yalin, Ishurd, O., Sun, Cuihong, and Jiang P., Structure analysis and antitumor activity of (1->3)- beta –D-glucans (Cordyglucans) from the mycelia of *C. sinensis. Planta Med.* 2005,**71**, 381-384.

42. Singh, R.P., Vinita Pachaujri, R.C., Verma, and Mishra, K.K., Caterpillar fungus (*Cordyceps sinensis*) -. A Review. *J. Eco-friendly Agriculture.* 2008,**3**, 1-15.

6

Ethnomycology and Ethnopharmacology of some edible and medicinal mushrooms

Varughese George*, Salees P. Abraham, Soniamol Joseph, T.P. Ijinu and
P. Pushpangadan

Amity Institute of Phytochemistry and Phytomedicine, 3-Ravi Nagar, Peroorkada
P.O., Thiruvananthapuram – 695 005, Kerala, India.
E-mail: georgedrv@yahoo.co.in

The relationship between man and mushrooms can be traced back to several millennia to the written manuscripts of ancient cultures of India, China etc. Mushrooms have attracted human attention owing to their brilliant colours, aroma and flavour. Many of the mushrooms have been used as delicacy in human food. Some mushrooms have attracted human attention because of their hallucinogenic properties. Such mushrooms have been used for a long time in religious and cultural practices of ancient communities. Though, mushrooms have been used in traditional medicine for a long time, their efficacy as drugs have been established only in recent years. Chemical and pharmacological investigations have unambiguously established the medicinal properties of several fungi which have been part of the ethno-medical practices of traditional communities from different parts of the world. In the present communication, ethnomycology and ethnopharmacology of a few fungi particularly *Ganoderma lucidum* have been discussed.

Introduction

Among the fungi, higher Basidiomycetes, especially mushrooms have been used as diet material for several millennia. They provide unlimited sources of therapeutically useful biologically active compounds. Mushrooms represent a major and as yet largely unexplored source of potent new pharmaceutical products. Of the approximately 14,000 known species, 2000 are safe for health and about 700 species are known to have significant pharmacological properties. The relationship between man and mushrooms can be traced far back to antiquity. The most fascinating aspect of mushroom usage is related to the psychoactive and hallucinogenic properties of some mushrooms. There is extensive literature implicating certain mushrooms in ancient religious rituals and practices. Hippocrates in the 5th Century BC recommended that cauterization should be made by means of fungus in order to cure certain complaints. Discorides (~200 AD) said that fungi could cure many complaints and were an almost universal remedy. In India, *Charaka Samhita*, the ancient medical treatise contains references to the use of mushrooms in the chapter on 'Annapana Vidhi', Charaka says

'*Sarpachatra varjyastu bahyo anyacatra*
Seeta peenasakartrica madhura gurvya evaca'.

According to Charaka mushrooms growing on decomposed bodies of snakes must be avoided and should not be consumed. Mushrooms have cooling property, are sweet in taste and produce stamina and body weight. Charaka classified mushrooms under Saka Varga (leafy vegetables). In Ayurveda mushrooms are considered as cardiotonic, as an effective remedy for diabetes, fever, asthma, cough, anorexia and as a cure for various dermatological conditions. In China, Fungi Pharmacopoeia drawn from ancient knowledge at least 2000 yeas old, describes more than 100 species of mushrooms used by practitioners of Traditional Chinese Medicine (TCM).

Medicinal mushrooms have established history of use in traditional ancient therapies. Contemporary research has validated and documented much of the ancient knowledge. Interdisciplinary research has demonstrated potent and unique properties of compounds extracted from a range of medicinal mushrooms in the last three decades. Modern clinical practices in Japan, China, Korea, Russia, and several other countries rely on mushroom derived preparations. In India, we have a long tradition of using mushrooms for food. The edibility of mushrooms was arrived through years of observations and experimentation. Even today several cases of mushroom poisoning are reported particularly due to the mistaken identity because of the fact that certain toxic species looks similar to the edible ones.

Ethno-mycological studies on some edible fungi from Kashmir Himalayas

During our mushroom forays in the Himalayan Forests we have observed that certain mushrooms are relished by monkeys. These are definitely suitable for human consumption. Some mushrooms are attacked by snails while they avoid many others. This also is an indication of the toxicity of certain species. Forest dwelling tribes are experts in identifying edible species. We have observed that certain species are their favourites. Some of the species collected by them from the Kashmir Forests are presented here.

Cantharellus cibarius

This is commonly known as the chanterelle, golden chanterelle or girolle (Figure 1). It is probably the best known species of the genus *Cantharellus*, if not the entire family of Cantharellaceae. It is orange or yellow, meaty and funnel-shaped. On the lower surface, underneath the smooth cap, it has gill-like ridges that run almost all the way down its stipe, which tapers down seamlessly from the cap. It has a fruity smell, reminiscent of apricots and a mildly peppery taste (hence its German name, Pfifferling) and is considered an excellent edible mushroom.

Fi. 1: *Cantharellus cibarius* (http://en.wikipedia.org/wiki/main_page)

This mushroom has been collected from the upper regions of Gulmarg Forests in Kashmir Himalayas. This is also reported from the upper regions of the Western Ghats in Kerala. This is a favourite mushroom of the nomadic tribe called Gujjars, who are traditionally shepherds. Chanterelles as a group are generally described as being rich in flavor, with a distinctive taste and aroma difficult to characterize. Some species have a fruity odour, others a more woody, earthy fragrance, and others still can even be considered spicy. The golden chanterelle is perhaps the most sought-after and flavorful chanterelle, and many chefs consider it on the same short list of gourmet fungi as truffles and morels. It therefore tends to command a high price in both restaurants and specialty stores in Europe.

There are many ways to cook chanterelles. Most of the flavorful compounds in chanterelles are fat-soluble, making them good mushrooms to sauté in butter, oil or cream. They also contain smaller amounts of water and alcohol soluble flavorings, which lend the mushrooms well to recipes involving wine or other cooking alcohols. Many popular methods of cooking chanterelles include them in sautés, soufflés, cream sauces, and soups. They are not typically eaten raw, as their rich and complex flavor is best released when cooked.

Boletus edulis

Boletus edulis, Fam. Boletaceae grows in deciduous and coniferous forests and tree plantations, forming symbiotic ectomycorrhizal associations with living trees by enveloping the tree's underground roots with sheaths of fungal tissue (Figure 2). The fungus produces spore-bearing fruit bodies above ground in summer and autumn. The fruit body has a large brown cap which on occasion can reach 35 cm (14 in) in diameter and 3 kg (6.6 lb) in weight. Like other boletes, it has tubes extending downward from the underside of the cap, rather than gills; spores escape at maturity through the tube openings, or pores. The pore surface of the *B. edulis* fruit body is whitish when young, but ages to a greenish-yellow. The stout stipe, or stem, is white or yellowish in colour, up to 25 cm (10 in) tall and 10 cm (3.9 in) thick, and partially covered with a raised network pattern, or reticulations.

Fig. 2: *Boletus edulis* (www.mushroomhobby.com)

Prized as an ingredient in various foods, *B. edulis* is an edible mushroom held in high regard in many cuisines, and is commonly prepared and eaten in soups, pasta, or risotto. The mushroom is low in fat and digestible carbohydrates, and high in protein, vitamins, minerals and dietary fibre. Although it is sold commercially, it is very difficult to cultivate. Available fresh in autumn in Central, Southern and Northern Europe, it is most often dried, packaged and distributed worldwide. Keeping its flavour after drying, it is then reconstituted and used in cooking. *B. edulis* is one of the few fungi sold pickled. The fungus also produces a variety of organic compounds with a diverse spectrum of biological activity, including the steroid derivative ergosterol, a sugar binding protein, antiviral compounds, antioxidants, and phytochelatins, which give the organism resistance to toxic heavy metals.

Coprinus comatus

Coprinus comatus, Fam. Agaricaceae, the shaggy ink cap, lawyer's wig, or shaggy mane, is a common fungus often seen growing on lawns, along gravel roads and waste areas (Figure 3). The young fruit bodies first appear as white cylinders emerging from the ground, then the bell-shaped caps open out. The caps are white, and covered with scales this is the origin of the common names of the fungus. The gills beneath the cap are white, then pink, then turn black and secrete a black liquid filled with spores (hence the "ink cap" name). This mushroom is unusual because it will turn black and dissolve itself in a matter of hours after being picked or depositing spores.

When young it is an excellent edible mushroom provided that it is eaten soon after being collected (it keeps very badly because of the autodigestion of its gills and cap). If long-term storage is desired, microwaving, sauteing or simmering until limp will allow the mushrooms to be stored in a refrigerator for several days or frozen. Processing must be done whether for eating or storage within four to six hours of harvest to prevent undesirable changes to the mushroom. The species is cultivated in China as food. The mushroom can sometimes be confused with the Magpie fungus which is poisonous.The agent responsible for unpleasant symptoms when consumed with alcohol, coprine, which is found in *Coprinopsis atramentaria*, has not been isolated from *C. comatus*. We have

Fig. 3: *Coprinus comatus*
(www.mykoweb.com)

collected this mushroom from the Willow plantations near the river from Gandharbal, Kashmir. Fresh mushrooms are sold in the markets of Srinagar, Kashmir during spring.

Clavaria vermicularis

Clavaria vermicularis, Fam. Clavariaceae is commonly known as fairy fingers and is found in Kashmir Himalayas (Figure 4). This has been collected from upper reaches of Gulmarg Forests in Kashmir. The mushroom is edible but watery and insubstantial. The Gujjar tribe who are traditionally nomadic shepherds collect this mushroom for food. An extract of the fruit bodies of *C. vermicularis* inhibited the growth of Sarcoma 180 and Ehrlich solid cancers in mice by 90% and 80%, respectively.

Fig. 4: *Clavaria vermicularis*
(www.herbarium.iastate.edu)

Ramaria botrytis

Ramaria botrytis, Fam. Gomphaceae, is an edible species, and some rate it as choice (Figure 5). The odor is not distinctive, while the taste is "slight", or "fruity". In Kashmir, the Gujjar tribe collects this mushroom and uses it as a vegetable along with Rotti made of Maize floor.

In the Gafagnana region of central Italy, the mushroom is stewed, or pickled in oil. However, one field guide rates the edibility as "questionable", warning of the possible danger of confusing specimens with the poisonous *Ramaria formosa*.

Fig. 5: *Ramaria botrytis* (www.herbarium.iastate.edu)

Fig. 6: *Morchella esculenta* (www.trakyamantar.com)

Fruit bodies contain the antioxidant protocatechuic acid. Extracts of the fruit body of *Ramaria botrytis* have been shown to favorably influence the growth and development of HeLa cells grown in tissue culture. The mushroom contains the chemical nicotianamine, an ACE (angiotensin-converting enzyme) inhibitor. Nicotianamine is known to be essential in iron metabolism and utilization in plants.

In a 2009 study of 16 Portuguese edible wild mushroom species, *R. botrytis* was shown to have the highest concentration of phenolic acids (356.7 mg per kg of fresh fruit body), made up largely of protocatechuic acid; it also had the highest antioxidant capacity. Phenolic compounds are common in fruits and vegetables and are being scientifically investigated for their potential health benefits associated with reduced risk of chronic and degenerative diseases.

Morchella esculenta

Morchella esculenta, Fam. Morchellaceae, is an economically important wild species (Figure 6). In the higher-altitude regions of Central and North-west Himalayas it is locally known as 'Guchhi'. Though *Morchella* as a genus is fairly easy to recognize, species differentiation within the genus is a difficult task. Six species, namely *M. esculenta, M. conica, M. deliciosa, M. angusticeps, M. arassipes and M. hybrida (M. semilibera)* have been reported from India. The fruit bodies of all the species of the genus are edible and are mainly used as a flavouring in soups and gravies. *M. esculenta* is an expensive product because of its rich nutritional value coupled with a unique flavour.

The local people cook ascocarps (the fruiting body) mixed with rice and vegetables, and consider it as nutritious as meat or fish. It is also used in health care, and medicinal purposes differ among traditional hill societies isolated by linguistic, cultural and terrain barriers. While the Bhotiya tribes (Central Himalaya) use a decoction of *M. esculenta* by boiling the fruiting bodies in water, local communities in the Kullu District of Himachal Pradesh (Western Himalaya) boil it in milk. Mushroom metabolites are also used as adaptogens and immunostimulants.

Ethnomycology in cultural practices

A significant archaeological documentation referring to old mushroom religious cult is located in Kerala (Figure 7). It belongs to the megalithic culture dating back to 1000 - 100 BC. The so called kuda-kallu (umbrella stone) near Trichur attracted the attention of ethno-mycologists. They resemble to an umbrella but more to a large mushroom. It is traditionally believed that these mushroom shaped stones are erected as tomb stones to mark the burial places.

Fig. 7: Megalithic "umbrella stones" (kuda-kallu) from Arryannor, Kerala, Southern India (1st Millennium B.C)

Mushrooms in religious rituals

It is believed that *Amanita muscaria*, the hallucinogenic fungus was a constituent of the Divine drink, Soma - which is reputed to give immortality (Figure 8). *A. muscaria* also had a prominent role in Mexican religious practices. The hallucinogenic effects of *A. muscaria* constituents have been now established. The major chemical constituents of *A. muscaria* are muscarine, muscimole, ibotenic acid and muscazone.

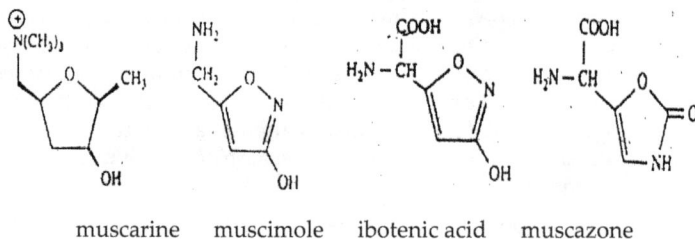

muscarine   muscimole   ibotenic acid   muscazone

Some medicinally important mushrooms

Some important medicinal mushrooms whose bioactivity has been established are listed below:

Agaricus blazei, Cordyceps sinensis, Flammulina velutipes, Ganoderma applanatum, Grifola frondosa, Ganoderma lucidum, Ganoderma oregonense, Hericium erinaceus, Inonotus obliquus, Lentinus edodes, Phellinus linteus, Pleurotus ostreatus, Polyporus umbellatus, Trametes versicolor etc.

Chemistry and pharmacology of *Ganoderma lucidum*

In Chinese folklore *Ganoderma lucidum,* Fam. Polyporaceae, has been regarded as a panacea for all types of diseases, perhaps owing to its demonstrated efficacy as a popular remedy to treat a large number of diseases (Figure 9). Traditionally this mushroom is widely used in the treatment of hepatopathy, chronic hepatitis, hypertension, arthritis, insomnia, bronchitis, asthma, gastric ulcers, hyperglycemia and cancer.

Ganoderma lucidum, is the best studied mushroom. There are more than 6500 publications on this fungus. Since it is an important medicinal fungus of China, most of the studies are carried out in China. It is used as a remedy for the treatment of number of ailments including cancer. Chemical investigation of this fungus have shown that it is prolific producer of novel mycochemicals. Recent investigations have resulted in the isolation of more than 200 triterpenoids from this mushroom. Among the various classes of chemical constituents isolated from this fungus are, Polysaccharides (β-1$\rightarrow$3 and β-1$\rightarrow$6-D glucans) which have shown antitumour properties, triterpenoids such as Ganoderic acids (more than 200 types of this lanostane type triterpenoids), sterols, proteins, nucleosides and fatty acids.

Fig. 8: *Amanita muscaria*

Fig. 9: *Ganode rma lucidum*

Ganoderic acid A   Ganoderic acid B   Ganoderic acid C

Ganoderic acid D   Ganoderic acid E   Ganoderic acid F

Ganoderic acid H   Ganoderic acid AM1

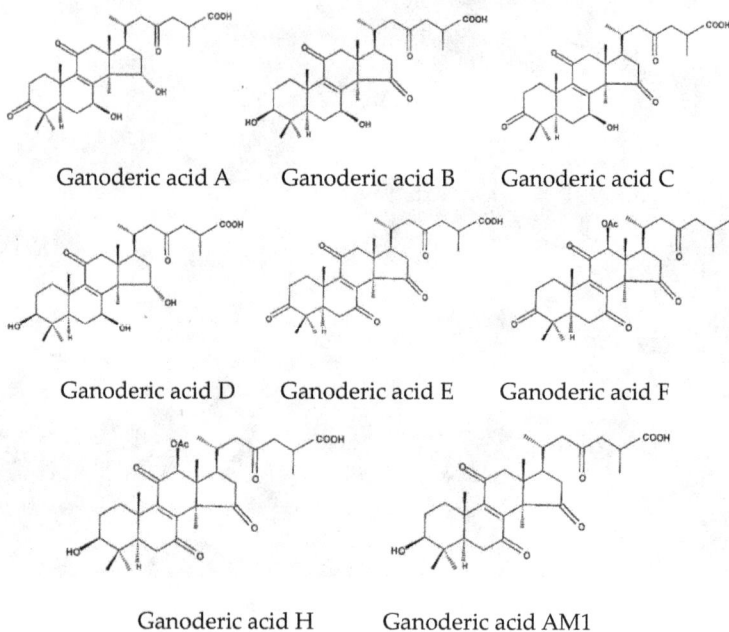

Recent findings from our laboratory

Our studies on *G. lucidum* collected from Thrissur, Kerala have resulted in the isolation of a novel epoxidic ganoderic acid, (8α,9α-epoxy-3,7,11,15,23-pentaoxo-5α-lanosta-26-oic acid (1), 3β-Hydroxy-7,11,15,23-tetraoxo-5α-lanosta-8-en-26-oic acid (2), ergosta-7,22-diene-3β-yl pentadecanoate (3), ergosta-7,22-diene-3β-ol (4), β-sitosterol (5), fatty acids (6-10), fatty acid ester (11) and octadecane (12).

Their structures were determined by 1H, 13C, 13C DEPT, 1H-1H COSY, HMBC, HSQC NMR, FT-IR, UV-Vis and FABMS spectral analysis. Compounds (1-3) exhibited good antifungal activity against *Candida albicans* in disc diffusion assay (100 μg/disc). Steroid ester (3) showed moderate anti-inflammatory activity (59.7% inhibition, 100 mg/kg) in carrageenan - induced paw edema.

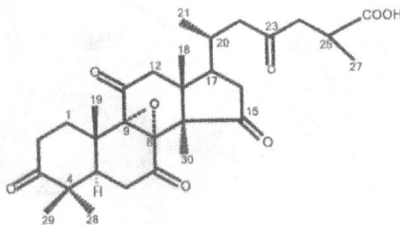

8α,9α-epoxy-3,7,11,15,23-pentaoxo-5α-lanosta-26-oic acid (1)[1]

Structure elucidation of 8α,9α-epoxy-3,7,11,15,23-pentaoxo-5α-lanosta-26-oic acid

Compound 1 was obtained as a pale yellow crystalline solid from the chloroform extract of the fruiting bodies of *G. lucidum*. Its Mass Spectrum showed the molecular ion at [M]+ m/z 528, corresponding to the formula $C_{30}H_{40}O_8$.

^{1}H NMR spectrum of 1 was analyzed with the1H-1H COSY data which showed signals for five tertiary methyls at δ 1.03 (3H, s), 1.06 (3H, s), 1.08 (3H, s), 1.28 (3H, s), 1.33 (3H, s), two secondary methyls at δ 0.94 (3H, d, J = 6.5 Hz), 1.22 (3H, d, J = 7.0 Hz), seven methylenes at δ 2.03/2.76, 2.06/2.64, 2.39/2.42, 2.46/2.48, 2.46/2.49, 2.55/2.81, 2.68/2.76 and four methine multiplets at δ 2.11, 2.18, 2.37 and 2.96.

The ^{13}C NMR and DEPT spectra showed the presence of 30 carbon atoms corresponding to seven methyls at δ 16.5, 16.9, 18.3, 19.6, 20.3, 21.0, 27.6, seven methylenes at δ 32.4, 33.2, 36.6, 41.1, 46.4, 47.9, 48.7, four methines at δ 31.6, 34.5, 43.4, 44.9, four quaternary carbons at δ 37.2, 45.9, 47.0, 55.3, two epoxidic quaternary carbons at δ 66.9, 67.1 and six carbonyl carbons including one carboxylic acid at δ 199.4, 201.9, 207.6, 208.3, 214.4 and 180.6.

Chemical shifts of the carbons in the ^{13}C NMR spectrum of 1 were identical to those of ganoderic acid E[2], except for C-8 and C-9. δ Values of these two carbons indicated an epoxy group between C-8 and C-9[3], instead of the olefinic linkage in ganoderic acid E. HMBC data of 1 showed correlation between δH 1.29 (H-19, s) to δC 67.1 and δH 1.33 (H-30, s) to δC 66.9. This confirmed C-8 and C-9 as the epoxy linked carbons in it. These spectral data proved 1 as 8α,9α-epoxy-3,7,11,15,23-pentaoxo-5α-lanosta-26-oic acid.

3β-hydroxy-7,11,15,23-tetraoxo-5α-lanosta-8-en-26-oic acid or ganoderic acid AM1 (2), has a double bond (C-8, C-9) and an -OH (C-3) instead of epoxide and carbonyl groups in 1. Ergosta-7,22-diene-3β-yl pentadecanoate (3) isolated both from the petroleum ether and chloroform extracts in this study was first reported from the n-hexane extract of *G. lucidum* procured from North America.

Compounds (4-12) were identified based on their spectroscopic data and comparison with literature as Octadecane, Dodecanoic acid, Tetradecanoic acid, Hexadecanoic acid, Octadecanoic acid, Methyl octadecanoate, Octadecadienoic acid, ß-Sitosterol, Ergosta-7,22-diene-3ß-ol [4-6].

Anti-inflammatory activity of 1-3

Compounds 1-3 were tested for anti-inflammatory activity using carrageenan-induced paw edema. The steroid ester, ergosta-7,22-diene-3β-yl pentadecanoate (3), showed 59.7% inhibition at 100 mg/kg body weight in carageenan-induced paw edema compared to 74.0% inhibition by the standard reference drug indomethacine at 10 mg/kg[7]

Antifungal activity of (1-3)

Antifungal activity of (1-3) were screened against the fungal strains viz., *Candida albicans* (Robin) Berkhout (MTCC 227), *Fusarium oxysporum* Schlecht (MTCC 284), *Aspergillus fumigatus* Fres (MTCC 343) and *A. niger* Tieghem (MTCC 1344) by the disc diffusion method. *C. albicans*, the most susceptible fungal strain, showed inhibition zones of 18 (1), 16 (2) and 22 mm (3) at 100 µg/disc compared to the antifungal agent, ketoconazole, which showed an inhibition zone of 30 mm at 100 µg/disc.

Bioactivity of fungal polysaccharides

Anti-tumor polysaccharides such as krestin, lentinan and schizophyllan have been isolated from various fungi. Polysaccharides from natural sources have diverse immunomodulatory activities both *in vitro* and *in vivo*. They are known modulators of cellular and humoral immunity. Polysaccharides derived from fungi are known for their antitumor and immunomodulating properties. They exert antitumor activity through activation of various immune responses in the host. Further, different strains of Basidiomycetes produce polysaccharides of varying properties.

Polysaccharides from *G. lucidum*

Polysaccharides isolated from *G. lucidum* and their antitumor and anti-inflammatory activities were investigated using in vivo models. Potential antitumor activity was shown by *G. lucidum* polysaccharides (GLP) against solid tumor induced by Ehrlich's ascites carcinoma cells. GLP at 100 mg/kg body mass showed 80.8 and 77.6 % reduction in tumour volume and tumour mass, respectively, when administered 24 h after tumour implantation. Again, GLP at the same dose but when administered prior to tumour inoculation, showed 79.5 and 81.2 % inhibition of tumour volume and tumour mass, respectively.

GLP showed significant dose-dependent activity in carrageenan-induced (acute) and formalin-induced (chronic) inflammation assays. At 100 mg/kg, GLP exhibited 57.6 and 58.2 % inhibition in carrageenan and formalin-induced assays, respectively[8].

References

1. Joseph, S., Janardhanan, K. K, George, V. and Sabulal, B.A., new epoxidic ganoderic acid and other phytoconstituents from *Ganoderma lucidum*, *Phtytochem Lett.* 2011,**4**, 386-388.

2. Komoda, Y., Nakamura, H., Ishihara, S., Uchida, M., Kohda, H. And Yamasaki, K., Structures of new triterpenoid constituents of *Ganoderma lucidum*, a polyporeceae, *Chem Pharmaceutic Bullet* 1985,**33**, 4829-4835.

3. Gonzalez, A. G., Leon, F., Rivera, A., Padron, J. I., Gonzalez-Plata, J., Zuluaga, J.C., Quintana, J., Estevez, F. and Bermejo, J., New lanostanoids from the fungus *Ganoderma concinna, J. Nat. Prod.* 2002,**65**, 417–421.

4. Lin, C.N., Tome, W.P. and Won, S.J., A lanostanoid of formosan *Ganoderma lucidum, Phytochem.* 1990,**29**, 673–675.

5. Rosecke J. and Konig W., Constituents of various wood-rotting basidiomycetes, *Phytochem.* 2000,**54**, 603–10.

6. Ziegenbein, F. C., Hanssen, H. P. and König, W. A., Secondary metabolites from *Ganoderma lucidum* and *Spongiporus leucomallellus, Phytochemistry* 2006,**67**, 202-211.

7. Joseph, S., Sabulal, B., George, V., Smina, T. P. and Janardhanan K. K., Antioxidative and antiinflammatory activities of the chloroform extract of *Ganoderma lucidum* found in South India, *Scientia Pharmaceutica*. 2009,77, 111-121.

8. Joseph, S., Sabulal, B., George, V., Antony, K. R. and Janardhanan, K. K., Antitumor and anti-inflammatory activities of polysaccharides isolated from *Ganoderma lucidum*, *Acta Pharmaceutica*. 2011, **61**, 335-342.

7

Ganoderma lucidum - A potential medicinal mushroom: It's cultivation and commercial production

A. K. Bordoloi*, Gitarthi Baruah and P.R. Bhattacharyya

CSIR-North East Institute of Science & Technology, Jorhat, Assam India

Ganoderma lucidum is called 'Ling Zhi' in Chinese, 'Reishi' in Japanese and 'young Zhi' in Korean, is one of the most famous medicinal mushrooms in the world. The fruit bodies have been used as the "Mushroom of immortality" in traditional Chinese medicine for more than 4000 years. During the last two decades, researchers have found various bioactive constituents. *G. lucidum* grows as a parasite on tree logs and stumps and causes serious harm to the host plant. Attempt has made to determine the yield performance of *G. lucidum* cultivated by using some easily available agro-wastes under control indoor agro-climatic condition of Jorhat. It is found that climatic condition of north-east is very favorable for the cultivation of *G. lucidum*.

Introduction

Ganoderma lucidum is a basidiomycetous fungus that grows on decaying logs and tree stumps. It is a polypore mushroom that is soft (when fresh), corky with a conspicuous red-varnished kidney-shaped cap. It lacks gills on its underside and releases its spores through fine pores. Initial fruiting body of *Ganoderma* is white, when mature it turns into yellow shades and finally into reddish to reddish brown shade, quite beautiful and distinctive. It has a worldwide distribution in both tropical and temperate geographical regions, including North and South America, Africa, Europe, and Asia, growing as a parasite or saprophyte on a wide variety of trees.

While talking about medicinal mushrooms, the name *G. lucidum* comes first to our minds. It is believed to be the oldest mushroom used in medicine and is one of the most respected medicinal mushrooms today. *G. lucidum* is called 'Ling Zhi' in Chinese, 'Reishi' in Japanese and 'young Zhi' in Korean, is one of the most famous medicinal mushrooms in the world[1]. The fruit bodies of *G.lucidum* have been used as the "Mushroom of immortality" in traditional Chinese medicine for more than 4000 years. Chinese scientists began investigating the chemical constituents and medicinal effects of *G. lucidum* in 1950s. Researchers from China, Japan, Korea and United States have formed an international *Ganoderma* research association, bringing its research to global platform. During the last two decades, researchers have found various bioactive constituents which have a wide range of actions, including boosting immune function, specifically natural killer cells, macrophage and interferon, as well as anti-viral and anti-bacterial; cardiovascular support, lowering blood pressure and serum total cholesterol and protecting the heart[2]. Research also demonstrated possible usefulness for allergies, bronchitis, cancer, HIV, inflammation, liver and protection from radiation[2].

It is an excellent herb which enables the body to heal itself. The Food & Drug Administration (FDA), USA, has categorized as a nutraceutical or a 'food & dietary supplement'. In India, it is now officially categorized as an Ayurvedic proprietary medicine or one which may be purchased over the counter without a prescription. Now a days, several *Ganoderma* extract based products are available in the market such as powder, coffee, soap, toothpaste, health drink, massage oil etc[3]. All these products contain *Ganoderma* extracts and are therefore good for health and fitness. In this present investigation, an attempt was made to determine the yield performance of *G. lucidum* cultivated by using some easily available agro-wastes under controlled indoor agro-climatic condition of Jorhat. Generally, *G. lucidum* grows as a parasite on tree logs and stumps and causes serious harm to the host plant so its cultivation in outdoor condition is not permissible.

Experimental

Location: The spawn production and the experiments were carried out at the NEIST, Jorhat, Assam.

Stock pure culture: Stock pure culture of *G. lucidum* obtained from Directorate of Mushroom Research, Indian Council of Agriculture Research, Solan, Himachal Pradesh was maintained on Potato Dextrose Agar (PDA) Media.

Spawn production: Rice grains were used for preparation of Spawn of *G. lucidum*. Grains were washed several times to remove any suspended particles. The grains were boiled in a container with water for 1/2 an hour. Boiled seeds were then dried under the sun. Twenty grams of Calcium carbonate were added to 1 kg of grain and mixed together properly in such a manner that every seed coated with a layer of calcium carbonate. Now these grains were filled into polypropylene bags (23cm long and 16 cm wide) upto ¾ of its volume. About 300 g of grains were packed in each bag.The bags were plugged with non absorbent cotton tightly and sterilized at 15 lbs per square inch steam pressure atleast for 15 to 20 minutes. After cooling, sterilized bags were inoculated with mycelial culture of *G.lucidum* in the Laminar air flow camber and incubated at 25±2⁰C. In order to obtain a good spawn, the bags were shaken carefully after two weeks of incubation in order to distribute the growing hyphae throughout the mixture. A good stage of maturation of *Ganoderma* spawn was obtained after 3-4 weeks of incubation.

Preparation of substrate: Substrate was prepared using saw dust (80%), wheat bran (18%), Sucrose (1%), Calcium carbonate (1%) on the dry wt basis of the substrate and were mixed thoroughly with water[4]. The correct water content was checked by pressing the medium by hand. The substrate was filled into polypropylene bags. About 300g of substrate was packed into each bag. Bags were plugged with cotton plug and autoclaved at 15 lbs per square inch pressure for 20 minutes.

After cooling, sterilized bags were inoculated with the paddy spawn. Each bag was inoculated with approximately 50 g of spawn under laminar air flow chamber and incubated in a dark chamber. Spawn run period completed 20-22 days later. When the mycelium had colonized on the substrate completely, bags were kept open in the cropping room at 30-32⁰C and 70-90% relative humidity. Water was sprinkled once in a day on the bags.

Mushroom Yield: The Cap formation of *G.lucidum* initiated in 2-5 days after opening the bags. Fruiting bodies were harvested when the caps become completely red and the white margin disappeared. Total yield (gm) was obtained from three flushes in a harvesting period of 45 days. Biological efficiency (BE) percentage ([fresh weight of harvested mushrooms/ dry matter content of the substrate] x100) was calculated[5,6].

The data given in the table 1 shows the yield performance of *G. lucidum* in three bags. Mushroom showed good yield on the substrate in all the three bags .In all cases approximately 40-50% of the total yield obtained from the first flush after which it goes on deceasing. About 64-80% BE of the substrate was recorded.

Table 1: Showing yield of *G. lucidum* in three flushes on synthetic logs:

No. of bags	Dry wt Of the Substrate in grams	SRP (days)	No of FlushesTotal yield BE (%) (weight of *G.lucidum* in grams) in gms				
1.	150	20	60.4	25.7	10	96.1	64
2.	150	22	70	40	10	120	80
3.	150	21	55.3	27.5	26	108.8	72.5

SRP = Spawn Run Period
BE =Biological efficiency

The yield performance in this experiment was promising under indoor agro-climatic condition of Jorhat, Assam and the substrate prepared by using agro wastes resulted in economic cultivation of *G. lucidum*.

With the plenty of raw materials, cheap labor and suitable environment with which the future cultivation of *Ganoderma* is very bright. A few suggestions are made here for increasing the yield for commercial cultivation under indoor control condition.

1. The indoor cultivation is to be practiced during cooler month of the year to get better yield.

2. Selection of right strain of *G. lucidum* which can give maximum yield under local indoor condition.

3. Testing of other agricultural wastes as substrate for production of *G. lucidum* in commercial scale.

Conclusion

It is found that climatic condition of north-east is very favorable for the cultivation of *G. lucidum* and one can easily cultivate it by using some locally available cheap agro-wastes. Medicinal mushrooms, value added products have a world trade of about $ 4 billion and till now it has been dominated by China, Japan and Korea. The fruit body of *G. lucidum* is sold in the market @Rs.600 to 700 per kg. Conservation and sustainable utilization of biological resources and economic enlistment of the people/ community inhabited within biodiversity hotspot is the prime concern worldwide. As demand of *G. lucidum* in world market is very high, there will be a good scope of taking up organized scientific cultivation of *G. lucidum* in large commercial scale. It is expected that the problem of unemployment will be solved to a great extend through this venture and upliftment of the rural economy of north-east, India.

Acknowledgement

Authors are thankful to the Director, NEIST, Jorhat for providing the necessary facilities to carry out the research work.

References

1. www.reishirescue.com

2. www.ganodrinker.com

3. www.ganoderma-for-health.com

4. Matute, R.G., Figlas, D., Devalis, R., Delmastro S. and Curvetto N., Sunflower seed hulls as a main nutrient source for cultivating *Ganoderma lucidum. Micologia Aplicada Int.* 2002; 14,19-24.

5. Nasreen, Z., Kausar, T., Nadeem, M. and Bajwa, R., Study of different growth parameters in *Ganoderma lucidum, Micologia Aplicada Int.* 2005,17,5-8.

6. Erkel. E.I., Yield performance of *Ganoderma lucidum* (Fr.) Karst cultivation on substrates containing different protein and carbohydrate sources. *African J Agri Res.* 2009,4,1331-1333.

8

Medicinal Mushrooms- scope for cultivation in Kerala

Lulu Das*and P.R. Prathibha

AICRP Mushrooms, Dept. of Plant Pathology, College of Agriculture, Vellayani, Triruvananthapuram, Kerala, India.

The nutritional value of mushrooms is noteworthy. Preparation of tasty recipes and standardization of value added products will encourage more people to consume these medicinal mushrooms. Around eighty different recipes including soup, shakes, main dishes, side dishes etc have been standardized at the College of Agriculture, Vellayani, Trivandrum. The best advantage is that all the medicinal mushrooms were obtained from Kerala soils naturally. Pink oyster mushroom (*Pleurotus eous*), giant mushroom (*Tricholoma giganteum*) and Jew's ear mushroom (*Auricularia* spp.) can be cultivated using paddy straw and sawdust

Introduction

More than 2000 species of fungi are reported to be edible throughout the world and about 283 of these are reported to be available in India. Out of these, only 20 mushroom species have been cultivated for edible purpose in different parts of the world. Kerala conditions favour the cultivation of a variety of crops thus providing tons of agro waste for cultivation of mushrooms. The need of the day is to exploit lesser known edible mushrooms perhaps the potent medicinal ones. Serious efforts must be taken to domesticate the lesser known species which have remained unexploited though suited for our tropical conditions. Priority should be given to those species which can be grown in wide range of temperature and can utilize the easily and amply available agricultural wastes in the State.

Medicinal mushroom varieties most suited for our climate are *Pleurotus eous*, *Pleurotus florida*, *Calocybe indica*, *Tricholoma giganteum*, *Auricularia polytricha* and *Lentinus* sp. The cultivation of all these varieties can be done successfully utilizing the agro wastes Oyster mushrooms are grown on paddywaste while *Auricularia* and *Lentinus* can be grown on sawdust. *Pleurotus* species are commonly known as Oyster mushrooms. They are edible tasty and constitute the second largest cultivated mushroom in the world. Oyster mushrooms are reported to possess several medicinal properties especially the hypocholesterolemic and antiproliferative activity [1]. *Pleurotus eous* belongs to Tricholomataceae. It is similar to *P. florida* in phytochemical constituents. *P.eous* has been found to have strong analgesic anti-inflammatory and antipyretic activity [2,3]. Crude extracts of *P. eous* also found to exhibit significant free radical scavenging property and antiplatelet activity, which is proved by the *in vitro* studies. *P. eous* has a dark pink colour which fades out when beds are kept in rooms with sufficient sunlight. An added advantage of this oyster mushroom is that it is most suited for cultivation in used mineral water bottles (Figure1). The pinheads appear within a short period of 10 days all over the surface of the bottles. Most suited for cultivation throughout the year in Kerala homesteads.

The recent and undoubtedly the most carefully executed experiments led to the development in stages of *Tricholoma*, a highly medicinal mushroom in the department of Plant Pathology. *Tricholoma giganteum*, commonly known as 'Giant Mushroom", found in nature after monsoon showers. It has a unique taste and is better suited for the tropical climate of Kerala. *Tricholoma* has a convex pileus, off-white to creamy white in colour, fleshy in texture and with a stout hairy stipe. *Tricholoma* is rich in protein, fibre, carbohydrates and vitamins and contains an abundant amount of essential amino acids and low in fat. Paddy straw is the most suited substrate for its cultivation and the steps are similar to that of milky mushroom bed laying. As in the case of milky mushroom, casing is important for emergence of pinheads. Vemicompost has been found to be the best casing material for growing *Tricholoma*.

Auricularia mushrooms are the fourth most important cultivated mushrooms in the world with a unique jelly taste. The common name Judas's ear comes from the legend that *Auricularia* formed its ear shaped fruiting bodies as a curse on the tree on which Judas hanged him self. The auriculariales are characterized by their typically brown colour and known for their tough rubbery consistency.

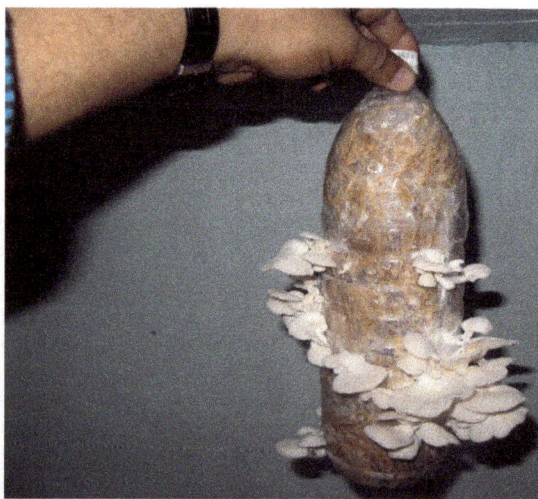

Fig. 1: Cultivation of oyster mushroom in used mineral water bottles

Besides its culinary value, *Auricularia* has significant medicinal properties and has been used for many centuries in traditional herbal remedies. According to the Doctrine of Signatures, a theory popular in Europe in the 1860's plants and fungi resembling certain parts of the body can be used to treat ailment of that part of the body. Since, the fungus resembles the folds of the throat *Auricularia* boiled in milk or vinegar was used to treat throat ailments. *Auricularia* has been particularly used to cure hemorrhoids and strengthen the body by stimulating the immune system. It was also used for treating widely varying conditions as spitting of blood, cardiac pain, diarrhea and against gastro intestinal upset. Fruiting bodies of *Auriculria* are rich in polysaccharides having antioxidant property [4]. Cultivation of *Auricularia* is done using sawdust [5]. Sawdust spawn has better shelf life than grain spawn [6]. Locally available cheap substrates that can be used for cultivation of *Auricularia* are the saw dust of rubber, anjili, mango, jack and coconut. Of these, rubber sawdust gave the maximum yield in three harvests. Food supplements like wheat bran and rice bran with paddy straw enhanced the yield of *Auricularia* but when paddy straw was used alone the yield was minimum [7]. Hence the locally available sawdust of different trees can be suitably utilized for production of this medicinal mushroom.

The nutritional value of mushrooms is noteworthy. Popularly known as the queen of vegetables it is ideal for those who prefer to stay slim because mushrooms provide us with lean protein as they have zero cholesterol fats and very low carbohydrates. Preparation of tasty recipes and standardization of value added products will encourage more people to consume these medicinal mushrooms. Around eighty different recipes including soup, snakes, main dishes, side dishes etc have been standardized at the College of Agriculture, Vellayani, Trivandrum [8]. Pink oyster mushroom (*Pleurotus eous*), giant mushroom (*Tricholoma giganteum*) and Jew's ear mushroom (*Auricularia* spp.) can be cultivated using paddy straw and sawdust. The best advantage is that all the medicinal mushrooms were obtained from Kerala soils naturally. The pink oyster mushroom has attracted many oyster mushroom growers in Kerala because of its keeping quality. The spawn of these two are sold in bulk to farmers indifferent parts of the State and also abroad. *Tricholoma* is perhaps the only mushroom to compete with the world button mushroom industry.

References

1.   Chang, Shu-Ting. and Philip M.G. *Mushrooms: cultivation nutritional value, medicinal effect and environmental impact.* CRC press, 1989, 4-6.

2.   Suseem, S.R., Saral, A.M., Reddy, P.N. and Gregory, M. Evaluation of analgesic activity of ethylacetate methanol and aqueous extracts of *Pleurotus eous* mushroom. *Asian Pacific Journal of Tropical Medicine* 2011, **4**, 117-120.

3.   Suseem, S.R., Saral A.M., Reddy, P. N. and Gregory M. Studies on antiflamatory and antipyretic activities of fruiting bodies of *Pleurotus eous* in experimental annuals. Pharmacology online. 2011, **1.**

4.   Zhang, S., Yu L. and Ma, L. Evaluation of antioxidant property and quality of breads containing *Auricularia auriculae* polysaccharide flour. *Food Chemistry.* 2006, **3,** 1158-116.

5.   Bhandal, and Mehta K.B. Cutivation of *Auricularia polytricha*Sacc (Jew.s ear mushroom). *Mushr Sci.* 1989,**12**, 20387-393.

6.   Prathibha, P.R. Standardization of techniques for cultivation of *Tricholoma giganteum* Massee in Kerala. M.Sc thesis, Kerala Agricultural University, 2013, pp.140.

7.   Vidya Resmi, C.R. Biology and cultivation of *Auricularia* sp M.Sc thesis, Kerala Agricultural University. 2008, pp.86.

8.   Das L. Cooking God,s own food-the south Indian way *Surrent vistas in mushroom biology and production.* 2003, pp.253-256.

9

Biodiversity of tropical Basidiomycetes from Konkan region and their biological properties

Shilpa A Verekar and Sunil K Deshmukh*

Department of Natural Products, Piramal Enterprises Limited, Mumbai 400 063
India. * email- sunil.deshmukh@piramal.com

Basidiomycetes are an important bio-resource of food and also used in traditional medicine all over the world. In developing countries, such as India and China alternative systems of medicine utilize the curative properties of basidiomycetes. Konkan region of the Western Ghats of India is a rich source of this group of fungi. These fungi are used by tribal's for medicinal purpose and also as food. During our survey for basidiomycetes for nutraceutical and pharmaceutical values from the konkan region the species of genera *Dictyophora, Clitopus, Tricholoma, Pleurotus, Auricularia, Clitocybe, Lepista, Oudmensiella, Lentinus, Volvariella, Termatomyces* were recorded. In this article the use of tropical basidiomycetes in traditional medicine/folk medicine, their distribution in the Konkan region (Mumbai, Thane and Raigad districts of Maharashtra) and medicinal properties of some of the mushrooms e.g. *Flammulina velutipes, Dictyophora indusiata, Pleurotus* sp. and *Tricholoma giganteum* are highlighted.

Introduction

Konkan, also called the Konkan Coast, is a rugged section of the western coastline of India from Raigad to Mangalore. Konkan division is also one of six administrative sub-divisions of the state of Maharashtra, comprising of its costal districts which include Maharashtra's districts of Raigad, Mumbai, Thane, Ratnagiri and Sindhudurg, the state of Goa, and Uttar Kannada, Udupi and Dakshina Kannada districts of Karnataka. The sapta-Konkan as depicted in Skanda-purana stretches from Maharashtra to Karnataka. Konkan division, rich in verdant natural beauty, is endowed with a coastline of 720 km with an area of 30746 sq. km. The main crops in this area are rice, nagli, mango, cashew, coconut, jackfruit, beetle-nut, spices.

Climate of Kokan region is generally hot and humid. The region witness all seasonal changes i.e. monsoon, winter and summer. Monsoon is generally from June to September, Konkan receives rainfall between 300 mm up to 900 mm. July and

August are the wettest months of the year. October to February, the temperatures are a bit milder with less humid conditions. Day temperatures are moderate and cool nights (15°C); an average temperature varies between 20 to 25°C. March to June is hot and humid. Hottest month is usually April. Average temperature for the summer season is between 32 to 40°C.

The general term Mushrooms is used for fungi which produce mainly large basidiomata or sometimes ascomata. The Indian subcontinent consists of several eco-climatic zones and therefore a treasure house of fungal diversity. In fact the "Western Ghats" which forms a long mountainous region along the west coast of India is considered as one of the hotspots of biodiversity. The first reference of occurrence of basidiomycetes from India was recorded by Montagne in 1842 .[1] Till date only about 1200 species of fungi belonging to the order Agaricales, Russulales and Boletales are described in comparison to about 14000 species of mushrooms reported from all over the world.[2-4] According to a recent estimate by Hawksworth [5] and Wasser and Didukh [6] this represents only 10% of the world's flora of mushrooms and the majority of the other 90% remains to be discovered in the tropical regions of the world. Mumbai, Thane and Raigad districts are rich in biodiversity and very limited attempts have been made for distribution of basidiomycetes. [4,7,8,]

Mushrooms are potential source of drugs either in extract form or as new chemical entities. For example Lentinan and PKS isolated from *Lentines edodes* are used as anticancer agents.[9] Betulinic acid isolated from *Inonotus obliquus* is the precursor of Betulin, effective against a variety of tumors.[10] Illudin S was isolated from various *Omphalotus* sp. like *O. olearius* and *O. nidiformis*. The derivative irofulven has a significantly superior therapeutic index in comparison to the original compound illudin S and possesses selectivity towards human tumor cells in which apoptosis is induced.[11]

Monacolin K (mevinolinic acid), an inhibitor of HMG CoA reductase was isolated from several *Pleurotus* sp. (e.g. *Pleurotus ostreatus*). Pleuromutilin an antibacterial compound was isolated from *Clitopilus passeckerianus* (formerly *Pleurotus passeckerianus*), and has also been found in *Clitopilus scyphoides*, and some other *Clitopilus* species.[12] Pleuromutilin and its derivatives are antibacterial drugs that inhibit protein synthesis in bacteria by binding to the peptidyl transferase component of the 50S subunit of ribosomes.[13] The derivatives of Pleuromutilin includes retapamulin (licenced for topical use in humans), valnemulin and tiamulin (approved for use in animals) and the investigational drugs azamulin and BC-3781 [14-16].

The present review deals with use of tropical basidiomycetes in traditional medicine/folk medicine, the distribution of edible species of mushrooms described from districts of Mumbai, Thane and Raigad of Konkan region of Maharashtra. The biological properties of some of the mushrooms including *Flammulina velutipes*, *Dictyophora indusiata*, *Pleurotus* sp. and *Tricholoma giganteum* are also covered.

Occurrence and Cultivation of Basidiomycetes

In tropical/ subtropical regions basidiomycetes occur in the monsoons generally after the first rains set in, till the end. We see these fungi growing on their natural habitat-dead, decayed or living plants, soil or dung. Cultures may be isolated either from fruiting bodies or basidiospores and maintained by subculturing, storage under mineral oil, in sterile water, and storage in liquid nitrogen. Short term storage on slants at 4°C is also possible. Lyophilization does not serve the same purpose.

Several difficulties are commonly encountered:

i) Seasonal occurrence of mushrooms and their collection in forests is difficult. They have a very short life and many are missed. To overcome these limitations, mushroom collectors have to visit the same locality time and again.

ii) Mushroom specimens, being fragile, are usually damaged while carrying them to the laboratory. In order to avoid damage during transportation, much care is required.

iii) Some of the mushrooms like those of *Coprinus, Dictyophora* etc. are deliquescent.

iv) Spores of many species e.g. from the genera *Inocybe* and *Russula* do not germinate.

v) Many mycelial cultures grow slowly on solid medium or in submerged cultures

vi) Fermentation cycles range from one week to five weeks or more.

A relatively small number of mushrooms are known because of the above constraints.

Distribution of basidiomycetes around Mumbai

A systemic survey was done to see the distribution of basidiomycetes in districts of Mumbai Thane and Raigad of Konkan region of Maharashtra since very little work has been recorded in this region. A field key to mushroom collection has been followed and used in describing the macro-and micro characteristics of gill fungi. [5,17-21] and some of the fungi were identified using various other keys. The results of the primary survey are given in Table 1 and Figure 1. 48 species of 18 genera were recorded and further work on identification is in progress.

Cytotoxic screening of metabolites produced by Basidiomycetes

Basidiomycetes produce a wide variety of secondary metabolites during their life cycle. These metabolites may be excreted into the growth medium or in the fruiting bodies when existing in the wild. Cytotoxic activity of some of the extracts are evaluated in Non-small cell lung cancer (NCI-H460), Human leukemia cells (HL-60), Colorectal carcinoma (HCT116), Breast cancer (MCF-7) and Pancreatic carcinoma (MIA PaCa-2) cell lines in comparison with Normal human fibroblast cells (WI-38). The mode of action of the active compounds are evaluated using high content screening system.[22]

Basidiomycetes in traditional medicine/ folk medicine

The invasion (migration) of the Aryans from central Asia into the Indian subcontinent took place around 1500 BC. They carried with them an intoxicating

Table 1: Species of various wild mushrooms genera recorded
from Konkan region

Sr. No.	Genus and Species	Habitat
1	*Agaricus arvensis* Schw . Fr	Solitary or scattered in lawns
2	*Agaricus brunnescens* Peck.	Solitary or scattered in lawns
3	*Agaricus essettei* Bon	Solitary on grassy soil
4	*Agaricus silvivola* (Vitt.) fr.	On dead wood and logs
5	*Armillaria mellea* (Fries) Kummer	On dead wood and logs
6	*Auricularia auricula* Judiae (L.) Berk	On dead wood and logs
7	*Auricularia polytricha* (Mont.) Sacc.	On dead wood and logs
8	*Cantharellus cibarius* Fr.	Associated with bamboo trees
9	*Lepista nuda* (Bull. : Fr.) Cooke	Solitary or scattered in lawns
10	*Clitopilus prunulus* (Scop.) P. Kumm.	Solitary or scattered in lawns
11	*Coprinus comatus* (Fr.) Gray	Singly to scattered on clustered in clumps on grassy land
12	*Crepidotus molis* (Schaelf. Fr.)Kummer	Growing on wood
13	*Crepidotus variabilis* (Pers.) P. Kumm.	Growing on wood
14	*Dictyophora indusiata* (Vent.: Pers.) Fischer	Solitary or scattered in lawns
15	*Fistulina hepatica* Fr.	On dead wood and logs
16	*Ganoderma applanatum* (Pers.) Pat.	On trunk of trees
17	*Ganoderma lucidum* (Leyss.: Fr.) Karst	Solitary or in groups on stumps
18	*Lentinus retinervis* Pegler,	Growing in single or in clusters in dead wood
19	*Lentinus squarossulus* Mont.	Growing in single or in clusters in dead wood
20	*Macrocybe gigantea* (Massee) Pegler & Lodge	Growing associated with Ficus benghalensies
21	*Macrolepiota procera* (Scop . Fr.) Sing	Solitary or scattered in lawns
22	*Macrolepiota rachodes* (Vitt.) Sing	Solitary or scattered in lawns
23	*Oudemansiella mucida* (Schaeff.Fr.) Hoehnel	Growing in soil along with fallen leaves
24	*Oudemansiella radicata* sensu Horak	Growing in association of unidentified forest tree
25	*Pleurotus columbinus* Quel	Growing in soil lawns/ grassy land

Contd.

Table 1 *Contd.*

26	*Pleurotus cytidiosus* Miller	Growing on dead tree in clusters.
27	*Pleurotus djamor* (Fr.) Boedeijn	Growing on dead tree in clusters.
28	*Pleurotus dryinus* (Pers.) P. Kumm.,	Growing on dead tree in clusters.
29	*Pleurotus eous* (Berk) Sacc.	Growing on dead tree in clusters.
30	*Pleurotus euosmus* (Berk. apud Hussey) Sacc.	Growing on dead tree in clusters.
31	*Pleurotus flabellatus* (Berk & Br.) Sacc.	Growing on dead tree in clusters.
32	*Pleurotus florida* Favose	Growing on dead tree in clusters.
33	*Pleurotus membranaceus* Mass.	Growing on dead tree in clusters.
34	*Pleurotus opuntiae* (Dur.and Lev.) Sacc.	Growing on dead tree in clusters.
35	*Pleurotus ostreatus* (Jacq.: Fr.)Kummer	Growing on dead tree in clusters.
36	*Pleurotus platypus* (Cook & Massee) Sacc.	Growing on dead tree in clusters.
37	*Pleurotus pulmonarius* (Fr.) Quel.	Growing on dead tree in clusters.
38	*Pleurotus sajor-caju* (Fr.) Sing.	Growing on dead tree in clusters.
39	*Pleurotus sapidus* Kalchor	
40	*Pleurotus spathulatus* (Fr.) Peck	Growing on dead tree in clusters.
41	*Termitomyces albuminosa* (Berk) Heim	On termite nests
42	*Termitomyces heimii* Natrajan	On termite nests
43	*Termitomyces mamiformis* Iteim	On termite nests
44	*Termitomyces microcarpus* (Berk& Br.) Heim	On termite nests
45	*Termitomyces robustus* (Beeli)Heim	On termite nests
46	*Volvariella bombycina* (Schaeff.) Singer,	Growing on ground
47	*Volvariella volvacea* (Bull. Fr.) Sing	Growing on rotten paddy straw
48	*Volvariella diplacea* Berk. & Br.	Growing on rotten paddy straw

drink **soma**. *Soma* was mostly used in Aryan religious rites. In the Rig Veda there are many songs on **soma**, only subject on which there are chapter in Rigveda. According to Wasson [23] the **soma** in the Rig Veda refers to *Amanita muscaria*. The use of mushrooms in traditional Chinese and Indian is well known. Lingzhi (*Ganoderma lucidum*) is a woody mushroom highly regarded in Chinese traditional medicine and is widely consumed in the belief that it promotes health and longevity, lowers the risk of cancer and heart disease and boosts the immune system.[24] Crude extracts of *Ganoderma tsugae* a traditional Chinese medicine, have been demonstrated to enhance splenic natural killer cell activity and serum interferon production in mice.[25] *Poria cocus* Wolf (Polyporaceae) is a well-known Chinese traditional medicine used for

Macrocybe giganteus   **Auricularia sp.**

Oudmensiella radicata   **Auricularia polytricha**

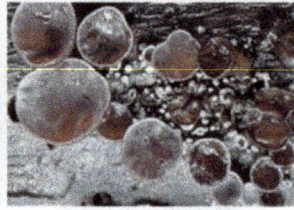

Volvariella volvaceae   **Pleurotus sp.**

Clitopilus prunulus   **Termitomyces sp.**

Pleurotus eous

Fig.1: Some of the common edible species from Konkan region

its diuretic, sedative, and tonic effects.[26] Inhibition of growth of influenza virus and recession of some kinds of cancer is shown by extract of Shiitake (*Lentinus edodes*). *Lentinus* contains compounds that reduce the serum cholesterol level in humans thereby lowering blood pressure. Wu Shui, a famous Chinese physician (Ming Dynasty) has written that the *Lentinus* is capable of curing cold, improving blood circulation, lowering blood pressure and generating stamina.[27]Many basidiomycetes have been used for medicinal purpose in India.[28-30] Some of the information on the mushrooms in folk medicine in India is given in Table 2.

Basidiomycetes as producers of bioactive metabolites

During our study we have observed compounds exhibiting significant anti-tumor activity along with other biological properties are summarized below. It must be noted that the basidiomycetes species listed here have been reported in India though the identified metabolites may be referred to from elsewhere.

Flammulina velutipes (Curtis) Singer

Flammulina velutipes (family Physalacriaceae) is an edible mushroom, commonly known as winter mushroom, velvet foot or enoki. This mushroom apart from its use as a delicacy has a large number of medicinal properties and some of them are highlighted here.

Flammulin, produced by the *F. velutipes* have antitumour properties.[31] Flammulin is a protein with a molecular weight of 40 kDa and is capable of inhibiting cell-free translation in a rabbit reticulocyte lysate system with an IC_{50} of 0.25 nM.[32] Shinnamin, a glycoproteins was also isolated from *F. velutipes* with antitumour activity.[33] Flammutoxin, a polypeptide of 22 or 32 kDa isolated from its fruiting bodies exhibit cardiotoxic and pore-forming cytolytic activity.[34]

In addition, a new antitumor glycoprotein "Proflamin" was found in mycelia of *F. velutipes*.[35] It is effective against allogeneic and syngeneic tumors by oral administration. Thus it is effective against solid sarcoma 180, B-16 melanoma, adenocarcinoma 775, and Gardner Lymphoma. It is also useful in combination therapy with other antitumor agents. Proflamin augments antibody formation and activates lymphocyte blastogenesis.[36]

A protein flammin (molecular mass of 30kDa), and velin (molecular mass of 19 kDa) were isolated from the fruiting bodies of *F. velutipes*. Flammin and velin inhibited translation in a rabbit reticulocyte lysate system with an IC_{50} of 1.4 and 2.5 nM, respectively.[37]

A single-chained ribosome inactivating protein velutin, with a molecular weight of 13.8 kDa was isolated from the fruiting bodies of the *F. velutipes*. It was capable of inhibiting human immunodeficiency virus (HIV-1) reverse transcriptase, β-glucosidase and β-glucuronidase.[38] Velutin, inhibited the expression of proinflammatory cytokines like TNF-α and IL-6 in low micromole levels by inhibiting NF-κB activation and p38 and JNK phosphorylation.[39] *Flammulina velutipes* sterol (FVS) consisted of mainly ergosterol (54.78%), 22,23-dihydroergosterol (27.94%) and ergost-8(14)-ene-3β-ol. FVS exhibited significant activity against U251 cells (IC_{50}=23.42µg/mL) while the HeLa cells were not significantly susceptible to

Table 2: Medicinal uses of Basidiomycetes in India

No.	Fungus	Vernacular Name	Medicinal Properties	References
1	*Amanita muscariai*		Used as a powder or tincture for swollen glands & epilepsy. Used in highly diluted preparations for heart ailment and rheumatoid arthritis.	Bahal 1994
2	*Amylosporus ampbellii (Polyporus anthelminticus)*	Bamboo Agaric	Anthelmintic	Chopara et al 1956
3	*Auricularia auricula*	Jew's ear	Used as poultice for inflammed eyes and as a gargle for inflammation of the throat.	Bahal 1994
4	*Astraeus hygrometrius*	Savan Putpura	Burn care	Sharma 1998
5	*Calvatia cyathiformis*	Dharti phool	Spore mass for wound healing and for checking pus formation	Sharma 1998
6	*Calvatia gigantea*		Used for anaesthesia	Bahal 1994
7	*Cythus limbatus*	Kulhari	For curing disorder of eyes like pain, redness or conjunctivitis	Sharma 1998
8	*Cythus stercoreus*	Nirgunthi	Conjunctivitis	Sharma 1998
9	*Daedaleopsis flavida*	Snuff Fungus	Snuff powder to reduce bilirubin and bilivirin for jaundice	Vaidya and Rabba 1993
10	*Fomes fomentarius*	Tinder Fungus	Cauterization of burned tissues.	Dymock et al 1890
11	*Fomes ignarius & Fomes fomentarius*		Used for rapid coagulation of blood	Bahal 1994
12	*Inonotus obliquus*	Chaga	Anticarcinogenic properties, chronic gastritis & ulcers	Rolf and Rolf 1925
13	*Larcifomes officinalis*	Larch - Quinine Fungus	Agarin ,agaricol and agaric acid is active principle. Liver Complaints, asthma, jaundice, dysentery, stomach pain, pain in joints,cathartic, lactifuge, diuretic, expectorants, check bleeding from bites.	Nadkarni 1954,
14	*Lentinus edodes*	Shiitake mushroom	Capable of generating stamina, curing cold, improving the blood circulation and lowering blood pressure. It stimulates the immune system which acts against cancer cells. It also has antiviral activity. Lowers cholesterol content in blood.	Pegler1983
15	*Lycoperdon giganteum*		Used as a soft and comfortable surgical dressing	Bahal 1994
16	*Lycoperdon pusilum*	Phusphus	Sporemass for controlling bleeding from cuts and also for wound heeling	Sharma 1998
17	*Meripilus giganteus*		Applied in gums to prevent excessive salivation, good styptic.	Khory 1887

Contd.

Table 2 *Contd.*

18	Microporus xanthopus	Saja Pihiri	For curing ear pain	Sharma 1998
19	*Phallus rubricandus*	Jhri Pihiri	For curing typhoid and for relief during labor pain.	Sharma 1998
20	*Phellinus grivus*		Conks used against kidney disorder	Vaidya and Rabba 1993
21	*Phellinus lintius*		Conks growing on bhendi to purify blood in skin diseases.	Vaidya and Rabba 1993
22	*Phellinus igniarius*	Bulgar tangali	Internally as a bitter tonic and laxative, externally as a styptic.	Nadkarni 1954, Chopara et al 1956
23	*Polyporus officinalis*	Agerick	Drastic purge external application to stop bleeding. Used for chronic catarrh diseases of the breast and lung, as a remedy for night sweating in tuberculosis, for rheumatism, gout, jaundice, dropsy and intestinal worms.	Bahal 1994
24	*Polyporus sp.*	Snuff Fungus	Narcotic snuff	Barkeley 1857
25	*Psilocybe sp.*		Hallucinogenic treatment. Treatment of mental disorders	Krauseman 1953
26	Pycnoporus sanguineus	Blood red mushroom	Dysentery, veneral disease, embrocation for leprous tubercies, inflamation of skin	Krauseman 1953
27	Termitomyces microcarpus	Bhoroan Pihiri	As a remedy in case of partial paralysis, tonic for over coming weakness.	Sharma 1998
28	Volvariella volvacea & Flammulina velutipes		They lower blood pressure & are active against tumor cells	Bahal 1994

the FVS. There activity was significantly enhanced in *F. velutipes* sterol nanomicelles (FVSNs) preparation than free drug.[40] A fungal immunomodulatory protein (FIP-fve), an activator of human T lymphocytes purified from *F. velutipes* has shown anti-tumor effect on oral administration in murine hepatoma model.[41] A stable hemagglutinin was isolated from the fruiting bodies of this mushroom, which inhibits proliferation of leukemia L1210 cells.[42]

Flammulinol A and flammulinolides A-G, as well as sterpuric acid, were isolated from the solid culture of *F. velutipes*. Flammulinolides A, B and F showed strong cytotoxicity against KB cell line with the IC_{50} of 3.9, 3.6, and 4.7 mM, respectively. Flammulinolide C showed strong cytotoxicity against Hela cell line with the IC_{50} of 3.0 mM.[43]

Enokipodins F -J, 2,5-cuparadiene-1,4-dione, enokipodin B and D, were isolated from the solid culture of *F. velutipes*. Compounds Enokipodin J, 2,5-cuparadiene-1,4-dione, enokipodin B and D, showed both moderate cytotoxicity against the human tumor cell lines (HepG2, MCF-7, SGC7901, and A549) and antioxidant activity in DPPH scavenging assay and compounds Enokipodin I, Enokipodin J, 2,5-cuparadiene-1,4-dione, Enokipodin B and D, displayed weak antibacterial activity against *Bacillus subtilis.* Enokipodin F, Enokipodin G and Enokipodin I showed weak antifungal activity against *Aspergillus fumigatus.*[44]

Water based extracts of *F. velutipes* was identified as novel anti-breast-cancer agents. It could markedly inhibit growth of (estrogen receptor) ER+ (MCF-7) and ER– (MDA-MB- 231) breast cancer cells. The extract induced an exceptionally rapid apoptosis on both types of cancer cells. The degree of cytotoxicity on ER– breast cancer cells was very high, whereas the ER– breast cancer cells are inhibited by about 99%, following FVE treatment.[45]

Hydrophilic extracts prepared from the fruiting body and spent culture medium of *F. velutipes* had total reducing power ability (RPA) and 2,2-diphenyl-1-picrylhydrazyl (DPPH) free radical scavenging activity (RSA), together with antioxidative activities against lipid oxidation in homogenates of yellowtail dark muscle and autoxidation of oxymyoglobin (oxyMb) purified from yellowtail dark muscle.[46]

Dictyophora indusiata (Vent.) Desv.

Dictyophora indusiata (Family Phallaceae) (Chinese name Zhu-Sun, the bamboo fungi) has been used as a medicinal mushroom to treat many inflammatory, gastric and neural diseases since 618 AD in China.[47] Polysaccharide extract from *D. indusiata* exhibited the antioxidant activities in various test models *in vitro* and *in vivo*, which in turn would influence the biological activity involving anti-inflammatory, immune enhancing and anticancer.[47-51] Even methanolic extract of the mushroom has excellent scavenging effects.[52]

A mannan (T-2-HN) and water-soluble glucans (T-4-N and T-5-N) from *D. indusiata,* were tested for antitumor activity against s.c. implanted sarcoma 180 in mice by i.p. administration. Considerable antitumor activity was observed with partially O-acetylated $(1\rightarrow 3)$-α-D-mannan (T-2-HN) at doses of 10 mg/kg/day x 10, water-soluble $(1 \rightarrow 3)$-β-D-glucans having β-1 $\rightarrow$ 6 linked D-glucosyl side chains (T-4-N and T-5-N) at doses 5 or 10 mg/kg/day x 10.[53] Triple-helix polysaccharide isolated from *D. indusiata* has significant inhibition action on S-180 tumor cells.[54]

Neuroprotective compounds, Dictyoquinazols A, B and C isolated from *D. indusiata* protected primary cultured mouse cortical neurons from glutamate and NMDA-induced excitotoxicities in a dose-dependent manner.[55] Two eudesmane-type sesquiterpenes, dictyophorines A and B were isolated from *D. indusiata*, which promoted nerve growth factor (NGF)-synthesis by astroglial cells.[56] A sesquiterpene albaflavenone was also isolated from *D. indusiata* having camphor-like odor which is active against *Bacillus subtilis.*[57]

The acid- and alkali-soluble polysaccharides (DIPs I and II) were extracted from the fruiting body of *D. indusiata*. The immunological assays showed that both DIP I and II had a noticeable effect on the hemolysis antibody level in the tested dosage range. However, DIP I could improve the weight of thymus organ and phagocytosis of monocyte. DIP II could restore delayed-type hypersensitivity reaction to dinitrofluorobenzene (DNFB), improving the activity of natural killer cells and the proliferation of splenocytes at high dose.[51]

Pleurotus sp.

Sixteen species (Familiy Pleurotaceae) has been found in this area. Some of the species of *Pleurotus* are cultivated in konkan region and consumed by locals. Here are some the medicinal properties of some of the *Pleurotus* sp. *Pleurotus florida* has antioxidant and antitumor activities in experimental animals.[58-59] Methanol extracts of *P. florida* inhibits inflammation and platelet aggregation. [60] Lavi et al. [61] reported that an aqueous polysaccharide extract from *Pleurotus ostreatus* induces anti-proliferative and pro-apoptotic effects on HT-29 colon cancer cells.

A novel water-soluble polysaccharide (POPS-1) obtained from the fruiting bodies of *P. ostreatus* exhibited significantly higher anti-tumor activity against HeLa tumor cell *in vitro*, in a dose-dependent manner, and exhibited significantly lower cytotoxicity to human embryo kidney 293T cells than HeLa tumor cells compared with anticancer drug 5-fluorouracil.[62] In their *in vitro* studies with *P. ostreatus* extracts against cancer cell (HL-60), the cytotoxic effect was reported due to presence of higher content of flavonoids in fruiting body. Cibacron blue affinity purified protein, protein fraction extracted from *P. ostreatus*, has been shown to have potent antitumor activity against different tumors using mice model.[63]

P. ostreatus ameliorates atherogenic lipid in hyper-cholesterolaemic rats,[64] possesses antitumor activity[65] and it has hypoglycaemic effects in experimentally induced diabetics and human subjects .[66-67] The ethanolic extract of *Pleurotus ostreatus* inhibited the cell proliferation in dose dependent manner, because after 72h of treatment with 0.1, 0.2 and 0.4 and 0.8 mg/ml of mushroom extract, inhibited the proliferation of HL-60 cells by 32.42, 47.2, 52.66 and 78.27 respectively. The extracts also exhibited a potent antioxidant activity against both *DPPH* and *ABTS* radicals screenings.[68]

Li et al. [69] isolated a homodimeric 32.4 kDa lectin from fresh fruiting bodies of the mushroom *Pleurotus citrinopileatus*. The lectin exerted potent anti-tumor activity in mice bearing sarcoma 180, and caused approximately 80% inhibition of tumor growth when administered intraperitoneally at 5 mg/kg daily for 20 days. Wong et al.[70] studied the *in vitro* anti-proliferative activities of the water-soluble polysaccharides extracted from the fruiting body and mycelium of *Pleurotus tuber-regium*. Fruiting body extract showed the strongest cytotoxicity (approximate IC_{50} 25 µg/mL) and exerted effective anti-proliferative activity at 200 µg/mL against human acute promyelocytic leukemia cells (HL-60). Both polysaccharide extracts induced apoptosis in HL-60 cells with an increase in the ratio of Bax/Bcl-2. Analysis from flow cytometry and western blot demonstrated that mycelium extract caused G2/M arrest in HL-60 cells by lowering the Cdk1 expression, while fruiting body caused S arrest in the HL-60 cells by a depletion of Cdk2 and an increase in cyclin E expression.

Ethanol extracts of *Pleurotus ferulae* have been shown to have antitumourigenic properties in human cervical cancer cell lines and in human lung cancer cell lines. When A549, SiHa and HeLa cells were incubated with different concentrations of ethanol extract of *P. ferulae*, the extracts showed strong cytotoxicity against A549 cells at concentrations over 10mg/mL and against SiHa and HeLa cells at over 40mg/mL.[71]

Pleurotus ostreatus (Oyster mushroom) has been shown to suppress proliferation of breast cancer (MCF-7, MDA-MB-231) and colon cancer (HT-29, HCT-116) cells, without affecting proliferation of epithelial mammary MCF-10A and normal colon FHC cells. Flow cytometry revealed that the inhibition of cell proliferation by *P. ostreatus* was associated with the cell cycle arrest at G0/G1 phase in MCF-7 and HT-29 cells. *P. ostreatus* also induced expression of the tumour suppressor p53 and cyclin-dependent kinase inhibitor p21(CIP1/WAF1). It appears that *P. ostreatus* suppresses the proliferation of breast and colon cancer cells via a p53-dependent as well as a p53-independent pathway.[72]

Treatment of mice with *Pleurotus ostreatus* at 100 and 500 mg/kg has suggested that *P. ostreatus* may prevent inflammation-associated colon carcinogenesis induced by 2-amino-1-methyl-6-phenylimidazo[4,5-b]pyridine (PhIP) and promoted by dextran sodium sulfate (DSS), via combined modulatory mechanisms of inflammation and tumor growth via suppression of COX-2, F4/80, Ki-67 and cyclin D1 expression in mice. However, incidence of colon tumors and high grade dysplasia was reduced by 50 and 63% only in the 500 mg/kg dose. [73]

A further study by the same group also showed that the anti-inflammatory activity of *Pleurotus ostreatus* is mediated through the inhibition of NF κB and AP-1 signaling. [74]An anti-proliferative effect of ethanol and water extracts of *Pleurotus tuberregium* against HCT-116 colon cancer cells has also recently been reported. [75]

Tricholoma giganteum Massee = *Macrocybe gigantea* (Massee) Pegler & Lodge

Tricholoma giganteum (*Macrocybe giganteum,* Family *Tricholomataceae*) is a very common mushroom found near *Ficus benghalensis* trees and growing on the ground in Konkan region of Maharashtra. The fruiting body of *T. giganteum* has many pharmaceutical uses and has long been utilized as a home remedy in Asia. An angiotensin I-converting enzyme (ACE) inhibitory peptide was isolated from *T. giganteum*. The maximum ACE inhibitory activity (IC_{50} 0.31 mg) was obtained when the fruiting body of *T. giganteum* was extracted with distilled water at 30° C for 3 h. The ACE inhibitory peptide was a novel tripeptide, showing very low similarity to other ACE inhibitory peptide sequences, and was sequenced as Gly-Glu-Pro. The purified ACE inhibitor from *T. giganteum* competitively inhibited ACE, and it maintained inhibitory activity even after incubation with proteases. ACE inhibitor from *T. giganteum* showed a clear antihypertensive effect in spontaneously hypertensive rats (SHR), at a dosage of 1 mg/kg.[76] Polysaccharide obtained from *T. giganteum* fruit bodies exhibited anticancer properties.[77-78] An anti-proliferative effect of clitocine from the mushroom *Leucopaxillus giganteus* on human cervical cancer HeLa cells has also been reported. The anti-proliferative effect was via an induction of apoptosis.[79] Polysaccharide TL-1, TL-2 and TL-3 were obtained from fruit body of *T. giganteum* and were screened for antioxidant activity. Polysaccharide TL-3, exhibited high activity for scavenging superoxide anion free radical and inhibiting mouse erythrocytic hemolysis (MEH) and malondialdehyde (MDA).[80] Antioxidant property of extracts obtained by different fractionation from *Tricholoma giganteum* basidiocarps was also reported by Chatterjee et al. [81] The ethanolic extract of *T. giganteum* exhibited hepatoprotective effect on carbon tetrachloride-induced hepatotoxicity in mice.[82] Ether extract of *T. giganteum* exhibited antimicrobial activity against *Salmonella* sp., *Escherichia coli, Bacillus thuringiensis, Staphylococcus aureus* and *Bacillus subtilis* at 2 mg/mL concentration. [83]

Conclusion

It is seen from this article that a good number of edible fungi occur in the Konkan region of Maharashtra. This gives an indication that these fungi should be explored as an alternative food for under privileged people of this region. The medicinal properties of some of these mushrooms clearly indicate that by consuming these mushrooms the chances of anti-tumor or immune-modulator diseases can be reduced. Most of these activities are due to presence of unusual polysaccharides present in them. The use of these mushrooms in traditional medicine is very significant as they can be a starting point for NCE research. If farmers are trained to cultivate these mushrooms then it will be great help to them as it will reduce the food burden and generate the revenue. There is a great need of a culture collection for mushrooms which can supply authentic basidiomycetes cultures as well as centres which can supply good quality spawn for cultivation.

References

1. Sathe, A.V., Agaricology in India–A review of work on Indian Agaricales. *Biovigyanam.*, 1979, **5**, 125 -130.

2. Manjula, B., A revised list of agrigicoid and boletoid basidiomycetes from India and Nepal. *Proc. Indian. Acad. Sci.* (Plant Sci.)., 1983, **92**, 81-213.

3. Natarajan K., Kumaresan V. and Narayanan K., A checklist of Indian Agaricales, and Boletales (1984 – 2002). *Kavaka.*, 2005, **33**, 61-128.

4. Deshmukh, S.K, Natrajan K. and Verekar, S.A., Poisonous and Hallucinogenic Mushrooms of India. *Int. J. Medicinal. Mushrooms.* 2006, **8:** 251-262.

5. Hawksworth, D.L., Mycologist's Handbook, CMI Kew Surrey England. 1974.

6. Wasser, S. P. and Didukh, M. Y., Mushroom polysaccharides in human health care. In: The Bio-Diversity of fungi - Their Role in Human Life" Deshmukh, S.K and Rai, M. K. Eds Science Publishers, Enfield, NJ. (2005). Pp

7. Bhide, V. P., Pande, A., Sathe, A. V., Rao, V. G. and Patwardhan, P. G., Fungi of Maharastra (Supp.1) M.A.C.S.Pub., 1987, pp 146.

8. Deshmukh, S.K., Biodiversity of tropical basidiomycetes - as sources of novel secondary metabolites. In "Microbiology and biotechnology in sustainable development" (ed. Jain P.C.) CBS Publishers & Distributors New Delhi India, 2004, pp. 116-135.

9. Ooi V.E. and Liu F., Immunomodulation and Anti-Cancer Activity of Polysaccharide-Protein Complexes . *Current Medicinal Chemistry.* 2000, 7, 715-729.

10. Mullauer, F.B., Kessler, J.H. and Medema, J.P., Betulinic acid, a natural compound with potent anticancer effects. *Anticancer Drugs.*, 2010, **21**, 215-227.

11. Schilder, R. J., Blessing, J. A., Shahin, M. S., Miller, D. S., Tewari, K. S., Muller, C. Y., Warshal, D., McMeekin, D. S. and Rotmensch J., A phase II evaluation of Irofulven (IND#55804, NSC#683863) as second-line treatment of recurrent

or persistent intermediately platinum-sensitive ovarian or primary peritoneal cancer: A Gynecologic Oncology Group trial. *Int. J. Gynecol. Cancer.*, 2010, **20**, 1137–1141.

12. Alarcon, J., Aguila, S., Arancibia-Avila, P., Fuentes, O., Zamorano-Ponce, E. and Margarita H., Production and purification of statins from *Pleurotus ostreatus* (Basidiomycetes) strains. *Zeitschrift Fur Naturforschung C*, 2003, **58**, 62-64.

13. Kilaru, S., Collins, C.M., Hartley, A.J., Bailey, A.M. and Foster, G.D., Establishing molecular tools for genetic manipulation of the pleuromutilin-producing fungus *Clitopilus passeckerianus. Appl. Environ. Microbiol.*, 2009, **75**, 7196-7204.

14. Long, K.S., Hansen, L.H., Jakobsen, L. and Vester, B., Interaction of pleuromutilin derivatives with the ribosomal peptidyl transferase center. *Antimicrob. Agents Chemother.*, 2006, **50**,1458-1462.

15. Lolk, L., Pøhlsgaard, J., Jepsen, A.S., Hansen, L.H., Nielsen, H., Steffansen, S.I., Sparving, L., Nielsen, A.B., Vester, B. and Nielsen, P., A click chemistry approach to pleuromutilin conjugates with nucleosides or acyclic nucleoside derivatives and their binding to the bacterial ribosome. *J. Med. Chem.*, 2008, **51**, 4957-4967.

16. Novak, R. and Shlaes, D,M., The pleuromutilin antibiotics: a new class for human use. *Curr. Opin. Investig. Drugs.*, 2010, **11**,182-191.

17. Smith, A.H. Mushroom and their natural habitats. Hafner Press New York. (1949)

18. Atkinson, G. F., Mushrooms edible, poisonous etc 2nd.Hafner Publishing Co. New York, 1961, pp 322.

19. Pegler, D. N., A preliminary agaric flora of East Africa . HMSO London., 1977, 615 pp

20. Singer, R. The Agaricales in Modern Taxonomy 4th ed. J Cramer Germany. (1986) pp. 912.

21. Atri, N. S. and Saini, S.S., Collection and study of Agarics –An Introduction *Indian J. Mush.*, 2000a, **18**: 1-5.

22. Periyasamy, G., Verekar, S.A., Khanna, A., Mishra, P.D. and Deshmukh, S. K., Anticancer activity of sclerotiorin, isolated from an endophytic fungus *Cephalotheca faveolata* Yaguchi, Nishim. & Udagawa. *Indian J. Exp. Biol.*, 2012, **50**, 464-468.

23. Wasson, G. R., Soma -Divine mushroom of immortality XIII, Heu Court Brace and world Inc. New York., 1969, pp 318.

24. Wachtel-Galor, S., Tomlinson, B. and Benzie, I.F., *Ganoderma lucidum* ("Lingzhi"), a Chinese medicinal mushroom: biomarker responses in a controlled human supplementation study. *Br. J. Nutr.*, 2004, **91**, 263-269.

25. Gan, K. H., Fann, Y. F., Hsu, S. H., Kou, K. W. and Lin, C. N., Mediation of the cytotoxicity of lanostanoids and steroids of *Ganoderma tsugae* through apoptosis and cell cycle. *J Nat. Prod.*, 1998, **61**, 485-487.

26. Cuellar, M.J., Giner, R.M., Recio, M.C., Just, M.J., Máñez, S. and Ríos, J.L. Two fungal lanostane derivatives as phospholipase A2 inhibitors. *J Nat Prod.*, 1996, **59**, 977–979.

27. Bahal, N., *Handbook on Mushrooms* IIIrd Edition Oxford and IBH Publishing Co. Pvt Ltd New Delhi Bombay Calcutta, 1994, PP 157.

28. Vaidhya, J.C. and Rabba, A. S., Fungi in Folk medicine. *The Mycologist.*, 1993, 7: 131-133.

29. Rai, B.K., Ayachi, S.S. and Rai A., A note note on ethno-myco-medicines from central India. *Mycologist.*, 1993, **7**, 192-193.

30. Sharma, N., Myco-myth, Mycetismus and Medicines. *Ethnobotany.*, 1998, **10**, 16-21.

31. Komatsu, N., Terakawa, H., Nakanishi, K. and Watanabe, Y., Flammulin, a basic protein of *Flammulina velutipes* with antitumor activities. *J Antibiot.*, 1963, **16**:139-143.

32. Wang, H.X. and Ng, T. B., Flammulin: a novel ribosome-inactivating protein from fruiting bodies of the winter mushroom *Flammulina velutipes*. *Biochem Cell Biol.*, 2000, **78**, 699-702.

33. Koichi, I. and Tetsuo, K., Preparation of carcinostatic substance Publication Number: JP 56-127317 A Publish Date: 06-Oct-1981

34. Tomita, T., Ishikawa, D., Noguchi, T., Katayama, E. and Hashimoto, Y., Assembly of flammutoxin, a cytolytic protein from the edible mushroom *Flammulina velutipes*, into a pore-forming ring-shaped oligomer on the target cell. *Biochem J.*, 1998, **333**, 129-137.

35. Ikekawa, T., Maruyama, H., Miyano, T., Okura, A., Sawasaki, Y., Naito, K. Kawamura, K. and Shiratori, K., Proflamin a new antitumor agent: preparation, physicochemcal properties and antitumor activity. *Japn. J. Cancer Res.*, 1985, **76**, 142-148.

36. Ikekawa, T., Enokitake *Flammulina velutipes* : Antitumor activity of extracts and polysaccharides. *Food Rev. Intern.*, 1995, **11**, 203-206.

37. Ng, T.B. and Wang, H.X., Flammin and velin: new ribosome inactivating polypeptides from the mushroom *Flammulina velutipes*. *Peptides.*, 2004, **25**, 929-933.

38. Wang, H. and Ng, T. B., Isolation and characterization of velutin, a novel low-molecular-weight ribosome-inactivating protein from winter mushroom (*Flammulina velutipes*) fruiting bodies. *Life Sciences.*, 2001, **68**, 2151–2158.

39. Xie, C., Kang, J., Li, Z., Schauss, A.G., Badger, T.M., Nagarajan, S., Wu, T. and Wu, X., The açaí flavonoid velutin is a potent anti-inflammatory agent: blockade of LPS-mediated TNF-α and IL-6 production through inhibiting NF-κB activation and MAPK pathway. *J. Nutr. Biochem.*, 2012, **23**, 1184-1191.

40. Yi, C., Sun, C., Tong, S., Cao, X., Feng, Y., Firempong, C.K., Jiang, X., Xu, X. and Yu, J. Cytotoxic effect of novel *Flammulina velutipes* sterols and its oral bioavailability via mixed micellar nanoformulation. *Int. J. Pharm.*, 2013, **448**, 44-50.

41. Chang, H.H, Hsieh, K.Y., Yeh, C.H., Tu, Y.P. and Sheu, F., Oral administration of an Enoki mushroom protein FVE activates innate and adaptive immunity and induces anti-tumor activity against murinehepatocellular carcinoma. *Int Immunopharmacol.*, 2010, **20**, 239–246.

42. Ng, T.B. and Ngai, P.H.K., An agglutinin with mitogenic and antiproliferative activities from the mushroom *Flammulina velutipes*. Mycologia. 2006, **98**, 167–171.

43. Wang, Y., Bao, L., Liu, D., Yang, X., Li, S., Gao, H., Yao, X., Wen, H. and Liu, H., Two new sesquiterpenes and six norsesquiterpenes from the solid culture of the edible mushroom *Flammulina velutipes*. *Tetrahedron*, 2012, **68**, 3012-30

44. Wang, Y., Bao, L., Yang, X., Li, L., Li, S., Gao, H., Yao, X. S., Wen, H. and Liu, H. W., Bioactive sesquiterpenoids from the solid culture of the edible mushroom *Flammulina velutipes* growing on cooked rice. *Food Chemistry.*, 2012b,**132**, 1346–1353.

45. Gu, Y.H. and Sivam, G., Cytotoxic effect of oyster mushroom *Pleurotus ostreatus* on human androgen independent prostate cancer PC-3 cells. *J. Med. Food.*, 2006, **9**, 196-204

46. Bao, H.N., Ochiai, Y. and Ohshima, T., Antioxidative activities of hydrophilic extracts prepared from the fruiting body and spent culture medium of *Flammulina velutipes*. *Bioresour Technol.*, 2010, **101**, 6248-6255.

47. Ker, Y.B., Chen, K.C., Peng, C. C., Hsieh, C.L. and Peng, R. Y., Structural Characteristics and Antioxidative Capability of the Soluble Polysaccharides Present in *Dictyophora indusiata* (Vent. Ex Pers.) Fish Phallaceae. *Evidence-based complementary and alternative medicine* : *eCAM*, 2011, 2011396013,

48. Gao, Q., Hua, Y. L., Zhao, M. M. and Tang, J., Optimization of the extraction of polysaccharides from *Dictyophora indusiata*. Xiandai Shipin Keji., 2010, **26**, 826-829.

49. Deng, C., Hu, Z., Fu, H., Hu, M., Xu, X. and Chen, J., Chemical analysis and antioxidant activity in vitro of a β-D-glucan isolated from *Dictyophora indusiata*. *Int. J. Biol. Macromol.*, 2012, **51**,70-75.

50. Li, X., Wang, Z., Wang, L., Walid, E. and Zhang, H., *In vitro* antioxidant and anti-proliferation activities of polysaccharides from various extracts of different mushrooms. *Int. J. Mol.Sci.*, 2012, **13**, 5801-5817.

51. Hua, Y., Gao, Q., Wen, L., Yang, B., Tang, J., You, L. and Zhao, M., Structural characterisation of acid- and alkali-soluble polysaccharides in the fruiting body of *Dictyophora indusiata* and their immunomodulatory activities. *Food Chemistry.*, 2012, **132**, 739-743.

52. Mau, J. L., Lin, H. C. and Chen C. C. Antioxidant Properties of Several Medicinal Mushrooms. *J. Agric. Food Chem.*, 2002, **50**, 6072-6077.

53. Ukai, S., Kiho, T., Hara, C., Morita M., Gao, A. and Naomi, H.Y., Polysaccharides in FungiXIII Antitumor activity of various polysaccharides isolated from *Dictyophora indusiata, Ganoderma japonicum,Cordyceps cicadae, Auricularia auricula-judae* and *Auricularia* Species. *Chem. Pharm. Bull.*, 1983, **31**, 741-744.

54. Chen, J., Wang, J. and Zheng, H., Preparation method of triple-helix polysaccharide of *Dictyophora indusiata*, and application to antitumor agent and immunostimulant. Faming Zhuanli Shenqing CN 101502532 A 20090812, 2009,

55. Lee, I.K., Yun, B.S., Han, G., Cho, D.H., Kim, Y.H. and Yoo, I.D., Dictyoquinazols A, B, and C, new neuroprotective compounds from the mushroom *Dictyophora indusiata*. *J. Nat. Prod.*, 2002, **65**, 1769-1772.

56. Kawagishi, H., Ishiyama, D., Mori, H., Sakamoto, H., Ishiguro, Y., Furukawa, S. and Li, J., Dictyophorines A and B, two stimulators of NGF-synthesis from the mushroom *Dictyophora indusiata*. *Phytochemistry.*, 1997, **45**, 1203-1205.

57. Huang, M., Chen, X., Tian, H., Sun, B. and Chen, H., Isolation and identification of antibiotic albaflavenone from *Dictyophora indusiata* (Vent:Pers.) Fischer. *J. Chem. Res.*, 2011, **35**, 659-660.

58. Nayana, J. and Janardhanan, K. K. Antioxidant and antitumour activity of *Pleurotus florida*. *Curr. Sci.*, 2000, **79**, 941-943.

59. Manpreet, K., Giridhar, S. and Khanna, P. K. In vitro and in vivo antioxidant potentials of *Pleurotus florida* in experimental animals. *Mushroom Res.*, 2004, **13**:21-26.

60. Nayana, J., Ajith, T.A. and Janardhanan, K.K., Methanol extract of the oyster mushroom, *Pleurotus florida*, inhibits inflammation and platelet aggregation. *Phytotherapy Research.*, 2004, **18**, 43-46.

61. Lavi, I., Friesem, D., Geresh, S., Hadar, Y. and Schwartz, B., An aqueous polysaccharide extract from the edible mushroom *Pleurotus ostreatus* induces anti-proliferative and pro-apoptotic effects on HT-29 colon cancer cells. *Cancer Lett.*, 2006, **244**, 61-70.

62. Tong, H., Xia, F., Feng, K., Sun, G., Gao, X., Sun, L., Jiang, R., Tian, D. and Sun, X., Structural characterization and in vitro antitumor activity of a novel polysaccharide isolated from the fruiting bodies of *Pleurotus ostreatus*. Bioresour Technol. 2009, **100**, 1682–1686.

63. Maiti, S., Mallick, S.K., Bhutia, S.K., Behera, B., Mandal M. and Maiti, T.K., Antitumor effect of culinary-medicinal oyster mushroom, *Pleurotus ostreatus* (Jacq.: Fr.) P. Kumm., derived protein fraction on tumor-bearing mice models. *Intl. J. Med. Mush.*, 2011, **13**, 427-440.

64. Hossain, S., Hashimoto, M., Choudhury, E.K., Alam, N., Hussain, S., Hasan, M., Choudhury, S.K. and Mahmud, I., Dietary mushroom (*Pleurotus ostreatus*) ameliorates atherogenic lipid in hyper- cholesterolaemic rats. *Clin. Exp. Pharmacol. Physiol.*, 2003, **30**, 470-475.

65. Yoshioka,Y., Tabeta, R., Saito, H., Uehera N. and Fukuoka, F., Antitumor polysaccharide from *P. ostereatus* (Fr.) Quel: isolation and structure of a beta glucan. *Carbohydr. Res.*, 1985, **140**, 93-100.

66. Chorvathoba, V., Bobek, P., Ginter, E. and Klvanova, J., Effect of the oyster fungus on glycemia and cholesterolemia in rats with insulin depended diabetes. *Physol. Res.*, 1993, **42**:175-179.

67. Khatun, K., Mahtab, H., Khanam, P.A., Sayeed, M.A. and Khan, K.A., Oyster mushroom reduced blood glucose and cholesterol in diabetic subjects. *Mymensingh Med J.*, 2007, **16**, 94-99.

68. Venkatakrishnan, V., Shenbhagaraman, R., Kaviyarasan, V., Gunasundari, D., Radhika, K. Dandapani, R. and Jagadish, L. K., Antioxidant and Antiproliferative Effect of *Pleurotus ostreatus*. *J. Phytol.*, 2010, 2, 022-028.

69. Li, Y.R., Liu, Q.H., Wang, H.X. and Ng, T.B., A novel lectin with potent antitumor, mitogenic and HIV-1 reverse transcriptase inhibitory activities from the edible mushroom *Pleurotus citrinopileatus*. *Biochim. Biophys. Acta.*, 2008, **1780**, 51-57.

70. Wong, S.M., Wong, K.K., Chiu, L.C.M. and Cheung, P.C.K., Non-starch polysaccharides from different developmental stages of *Pleurotus tuber-reginum* inhibited the growth of human acute promyelocytic leukemia HL-60 cells by cell-cycle arrest and/or apoptotic induction. *Carbohydr Polym.* 2007, **68**, 206–217.

71. Choi, D.B., Cha, W.S., Kang, S.H. and Lee, B.R. Effect of *Pleurotus ferulae* extracts on viability of human lung cancer and cervical cancer cell lines. *Biotechnol. Bioprocess Eng.*, 2004, **9**, 356-361.

72. Jedinak, A. and Sliva, D. *Pleurotus ostreatus* inhibits proliferation of human breast and colon cancer cells through p53-dependent as well as p53-independent pathway. *Int. J. Oncol.* 2008, **33**, 1307-13.

73. Jedinak, A., Dudhgaonkar, S., Jiang, J., Sandusky, G. and Sliva, D., *Pleurotus ostreatus* inhibits colitis-related colon carcinogenesis in mice. *Int. J. Mol. Med.*, 2010, **26**, 643-650.

74. Jedinak, A., Dudhgaonkar, S., Wu, Q.L., Simon, J. and Sliva, D., Anti-inflammatory activity of edible oyster mushroom is mediated through the inhibition of NF-κB and AP-1 signaling. *Nutr J.*, 2011, **10**, 52.

75. Maness, L., Sneed, N., Hardy, B., Yu, J., Ahmedna, M. and Goktepe, I., Anti-proliferative effect of Pleurotus tuberregium against colon and cervical cancer cells. *J. Med. Plants Res.*, 2011, **5**, 6650–6655.

76. Hyoung, L. D., Ho, K.J., Sik, P. J., Jun C. Y. and Soo, L. J., Isolation and characterization of a novel angiotensin I-converting enzyme inhibitory peptide derived from the edible mushroom *Tricholoma giganteum*. Peptides., 2004, **25**, 621-627.

77. Mizuno, T., Kinoshita, T., Zhuang, C., Ito, H. and Mayuzumi, Y, Antitumor-active heteroglycans from Niohshimeji mushroom, *Tricholoma giganteum*. Biosci. Biotechnol. Biochem., 1995, **59**, 568– 571

78. Mizuno, T., Yeohlui, P., Kinoshita, T., Zhuang, C., Ito, H. and Mayuzumi, Y. Antitumor activity and chemical modification of polysaccharides from Niohshimeji mushroom, *Tricholoma giganteum. Biosci. Biotechnol. Biochem.*, 1996, **60**,30–33

79. Ren, G., Zhao, Y.P., Yang, L. and Fu, C.X., Anti-proliferative effect of clitocine from the mushroom *Leucopaxillus giganteus* on human cervical cancer HeLa cells by inducing apoptosis. *Cancer Letters.*, 2008, **262**, 190–200.

80. Chen, Y., Wang, C., Wang, R., Hu, M., Duan, P. and Zhai, T., Method for preparation of natural antioxidant *Ttricholoma giganteum* polysaccharide. Faming Zhuanli Shenqing CN 102603908 A 20120725, 2012,

81. Chatterjee S., Saha G.K. and Acharya K. Antioxidant activities of extracts obtained by different fractionation from *Tricholoma giganteum* basidiocarps, *Pharmacologyonline*, 2011, **3**, 88-97.

82. Acharya, K., Chatterjee, S., Biswas, G., Chatterjee, A. and Saha, G. K. Hepatoprotective effect of a wild edible mushroom on carbon tetrachloride-induced hepatotoxicity in mice. Int. *J. Pharm. Pharm. Sci.*, 2012, **4**, 285-288.

83. Mo, M. and Zhang, Q., Studies on antimicrobial activity of extracts of *Macrocybe gigantean. Shipin. Gongye. Keji.*, 2009, **30**, 151-153, 161.

10

Recent developments in the prospects of mushrooms as nephroprotective agent: A review

B. Nitha[1] and K.K Janardhanan*[2]

[1] Sree Ayyappa College, Eramallikkara,Alappuzha-689109, Kerala. India,
[2] Amala Cancer Research Centre, Thrissur-680555, Kerala, India.
Email: *kkjanrdhanan@yahoo.com

From time immemorial, mushrooms have been valued by humankind as a culinary wonder and folk medicine in Oriental practice. The last decade has witnessed the overwhelming interest of western research fraternity in pharmaceutical potential of mushrooms. The chief medicinal uses of mushrooms discovered so far are as anti-oxidant, anti-diabetic, hypocholesterolemic, anti-tumor, anti-cancer, immunomodulatory, anti-allergic, nephroprotective, and anti-microbial agents. The mushrooms credited with success in treating kidney damages belong to the genus *Phellinus, Pleurotus, Ganoderma, Morchella* and *Grifola* The present review updates the recent findings on the pharmacologically active compounds, their nephroprotecive effects, and underlying mechanism of biological action in order to raise awareness for further investigations to develop nephroprotecting therapeutics from mushrooms. The mounting evidences from various research groups across the globe, regarding the nephroprotective application of mushroom extracts unarguably make it a fast-track research area worth mass attention.

Introduction

From time immemorial, mushrooms have been valued by humankind as a culinary wonder and folk medicine in Oriental practice. The last decade has witnessed the overwhelming interest of western research fraternity in pharmaceutical potential of mushrooms. The chief medicinal uses of mushrooms discovered so far are as anti-oxidant, anti-diabetic, hypocholesterolemic, anti-tumor, anti-cancer, immunomodulatory, anti-allergic, nephroprotective, and anti-microbial agents. The mushrooms credited with success against cancer belong to the genus *Phellinus, Pleurotus, Ganoderma, Morchella* and *Grifola*. The present review updates the findings on the nephroprotective effect of mushrooms. The mounting evidences from various research groups across the globe, regarding various medicinal properties of mushroom extracts unarguably make it a fast-track research area worth mass attention. Mushrooms have been regarded as gourmet cuisine across the globe since antiquity for their unique taste and subtle flavor. Recently, it has been discovered that many mushroom species are miniature pharmaceutical factories producing hundreds of novel constituents with miraculous biological properties. They have a long history of use in Oriental medicine, but their legendary effects in promotion of good health and vitality are being supported by contemporary studies only. Of late, mushrooms have emerged as wonderful source of nutraceuticals, anti-oxidants, anti-cancer, prebiotic, immunomodulating, nephroprotective, hepatoprotective, anti-inflammatory, cardiovascular, anti-microbial, and anti-diabetic [1, 2, 3, 4, 5]. The ongoing research projects are aimed to promote mushrooms as new generation "biotherapeutics".

Recently, a number of bioactive molecules have been identified from various mushrooms. The bioactive compounds of mushrooms include polysaccharides, proteins, fats, ash, glycosides, alkaloids, volatile oils, tocopherols, phenolics, flavonoids, carotenoids, folates, ascorbic acid enzymes, and organic acids. Several active components from mushrooms such as lentinan, krestin, hispolon, lectin, calcaelin, illudin S, psilocybin, *Hericium* polysaccharide A and B (HPA and HPB), ganoderic acid, schizophyllan, laccase, etc have been isolated till recently. Polysaccharides are the best known and most potent mushroom-derived substances with biological properties. The polysaccharide, β-glucan is the most versatile metabolite due to its broad spectrum biological activity. These β-glucans consist of a backbone of glucose residues linked by β-(1 → 3)-glycosidic bonds, often with attached side-chain glucose residues joined by β-(1 → 6) linkages [6]. Their mechanisms of action involve is recognized as non-self molecules, so the immune system is stimulated by their presence. The scientific investigations to support the claims have gained momentum in recent years.

The kidneys are dynamic organs and represent the major control system maintaining body homeostasis, affected by many chemicals and drugs. The kidneys have important physiological functions including maintenance of water and electrolyte balance, synthesis, metabolism and secretion of hormones, and excretion of the waste products from metabolism. In addition, the kidneys play a major role in the excretion of drugs, hormones, and xenobiotics[7].

A number of cancer chemotherapeutic agents and drugs of either natural or synthetic origin have been developed during recent years. Drug- induced kidney disease constitutes an important cause of acute renal failure and chronic kidney disese in present day clinical practice. The incidence of drug-induced nephrotoxicity has been increasing with the ever increasing number of drugs and with easy availability of over- counter medication viz- non asteroidal anti-inflammatory drugs (NSAIDs). Antibiotics, NSAIDS, chemotherapeutic agents, angiotensin converting enzyme inhibitors are the major culprit drugs contributing to kidney damage. Drug induced acute renal failure (ARF) account for 20% of all ARF in an Indian society of which aminoglycosides account for 40% of total cases[8].

Mushrooms as nephroprotective agents

Phellinus

Phellinus is a genus of mushrooms belonging to the family Hymenochaetaceae. The dominant and most frequently found species are *Phellinus (P. senex), P. rimosus, P. badius, P. fastuosus, P. adamantinus, P. caryophylli* and *P. durrissimus* [9]. About 18 species are found to occur in Kerala, most of them are wood inhabiting[10]. *Phellinus romosus* is a wood inhabiting poly pore macro fungus generally seen on jack fruit tree trunks in Kerala. Ajith *etal* reported the nephroprotective effect of Phellinus rimosus (Berk) Pilat . Pretreatment of ethyl acetate extract of *Phellinus rimosus* (Berk) Pilat protected cisplatin induced nephrotoxicity in mice. Single dose of the extract (25 and 50 mg/kg body weight) decreased cisplatin induced serum creatinine and urea levels. The cisplatin depleted renal antioxidant defence system, such as activity of superoxide dismutase (SOD), catalase (CAT), glutathione peroxidase (GPX) and concentration of reduced glutathione (GSH), were restored by the treatment of the extract. The enhanced renal antioxidant defence system prevented cisplatin induced tissue lipid peroxidation. The nephroprotective effect of *P. rimosus* may possibly by enhancing renal antioxidant status[11].

Ganoderma

Ganoderma, commonly known as Lingzhi or Reishi, also called mushroom of immortality, belonging to family Ganodermataceae has been traditionally administered throughout Asia for centuries for the treatment of many ailments. *G. lucidum* is a well-known Asian herbal remedy with a long and impressive range of applications. Global consumption of *G. lucidum* is high, and a large, increasing series of patented and commercially available products that incorporate *G. lucidum* as an active ingredient are available as food supplements. The methanolic extract of *G.lucidum* was reported to possess significant nephroprotective effect aginst the cisplantin induced toxicity. The administration of the extract restored the serum urea, creatinine, the levels renal antioxidants superoxide dismutase (SOD), catalase (CAT),glutathione peroxidase (GPx), reduced glutathione (GSH) and also inhibited the lipid peroxidation in animals with cisplatin induced renal damage[12,13]. Ying-Hua Sheih *et al* reported the renal protective effect of *Ganoderma lucidum* in mice against the damage induced by ethanol administration. The renal homogenates showed a higher level of malon di aldehyde in ethanol administered group. In the animals treated with *Ganoderma lucidum* the malon di aldehyde level was found to be lowered. The study conclude that the renal protective effect might be due to at least in part to its prominent superoxide scavenging effect and *Ganoderma* can protect kidneys from superoxide induced renal damage [14].

Pleurotus

Oyster mushroom (*Pleurotus* species) are excellently edible and nutritious, rank among one of the most widely cultivated mushrooms in the world[15]. *Pleurotus* species have high medicinal value. Compounds extracted from these mushrooms exhibit activity against various chronic diseases. The medicinal beneficial effects of *Pleurotus* species were discovered independently in different countries. The awareness of their medicinal properties came not only from Asia but from the folklore of Central Europe, South America and Africa [16].

Pleurotus eous is an edible oyster mushroom currently available in Southern part of India. A study by Sasikumar and Sudha indicate that the water extract of *Pleurotus eous* possesses profound nephroprotective activity against cisplatin induced kidney damage. In the study the cisplatin induced animals showed elevated levels of urea, creatinine, uric acid coupled with decreased levels of protein. Treatment with two different doses (500 and 1000 mg/kg b.wt.) of *Pleurotus eous* altered all the biochemical constituents to near normal levels indicating its significant nephroprotective effects. The study reports that cisplatin administration produced a two fold increase in the levels of thiobarbituric reactive substances (TBARS) and hydroperoxides, as compared to control rats. The levels of the TBARS and hydroperoxides were significantly reversed on treatment with the *Pleurotus eous* in a dose dependent manner. The activities renal antioxidants enzyme activities such as CAT, SOD and GPx, GST was decreased significantly in the cisplatin treated group of animals compared to normal group. Treatment with *Pleurotus eous* showed restoration of all the enzyme activities to near normal. The activity elicited by the extract might be due to its ability to activate antioxidant enzymes [17].

Hajizadeh Moghaddam *et al* reported the nephroprotective activity of *Pleurotus porrigens* against gentamicin induced kidney damage [18]. The ethyl acetate extract of *Pleurotus porrigens* restored the level of serum urea, creatinine and blood urea nitrogen level almost to the normal level in gentamicin treated mice. The possible mechanism of nephroprotective effect of this mushroom may be attributed to its significant antioxidant and free radical scavenging activity elicited by the phyto chemical compounds such as phenols and flavanoids.

Jayakumar *et al.*, carried out a study for evaluating the protective effect of the oyster mushroom, *Pleurotus ostreatus* on carbon tetrachloride (CCl_4)-induced renal toxicity in male Wistar rats. *P. ostreatus* extract was used to treat rats with CCl_4-induced renal damage. The study reports that administration of *P. ostreatus* lowered the mean level of MDA, elevated the mean levels of GSH and of vitamins C and E and enhanced the mean activities of CAT, SOD, Gpx and GST so that the values of most of these parameters did not differ significantly from those of normal rats. Histopathological studies confirmed the toxic effects of CCl_4 on kidneys, and also tissue protective effect of the extract of *P. ostreatus*. These results suggest that an extract of *P. ostreatus* is able to alleviate the oxidative damage caused by CCl_4 in the kidneys of Wistar rats [19].

A study undertaken to investigate the putative protective effect with antioxidant potential of the oyster mushroom (*Pleurotus ostreatus*) in glycerol-induced ARF in rats by Sirag [20]. Oxidative stress markers in kidney, kidney abnormalities as well as hematological alteration were estimated. Glycerol treatment resulted in a marked renal oxidative stress, a significantly deranged renal function and a reduction in all hematological tested parameters. Pre-treatment with mushroom markedly reduced elevated thiobarbituric acid reacting substances, reduced glutathione, protein carbonyl and resorted the depleted renal antioxidant enzymes, attenuated renal dysfunction and the disturbances observed in hematological and biochemical parameters. These results suggest that *Pleurotus ostreatus* may have ability to protect the renal damage involved in acute renal failure in rats.

Nuhu Alam et al evaluated the comparative effects of oyster mushrooms on liver and kidney function in hyper and non cholesterolemic rats. Feeding of hyper cholesterolemic rats a 5% powder oof oyster mushrooms (*Pleurotus ostreatus, P. sajorr- caju and P.florida*) reduced the plasma total cholesterol level and triglyceride level. How ever, it had no adverse effect on plasma creatinine and urea nitrogen level. The study revealed that feeding of 5% oyster mushroom powder does not have detrimental effects on kidneys [21].

Morchella

Members of the genus *Morchella* (Morchellaceae), commonly known as morels, are among the most highly priced edible mushrooms in the world. They are characterized by a hollow fruit body consisting of a pitted cap with an intergrown stipe[22]. Morels are locally known as *Guchhi* and are used in health care and medicinal purposes among traditional hill societies[23]. *Morchella esculenta* (L) Pers. is an excellently edible and nutritious morel mushroom belonging to Ascomycotina division. In India, this mushroom is found growing in the forests of Jammu and Kashmir and Himachal Pradesh. The aqueous–ethanol extract of *Morchella esculenta* is reported to possess significant nephroprotective activity against cisplatin and gentamicin induced kidney damages[24]. Treatment with the extract at 250 and 500mg/kg body weight decreased the cisplatin and gentamicin induced increase in serum creatinine and urea levels. Administration of the extract restored the depleted antioxidant defense system. The decreased activity of superoxide dismutase (SOD), catalase (CAT), glutathione peroxidase (GPx), and reduced glutathione (GSH) in the kidneys consequent to cisplatin and gentamicin administration was significantly elevated. The treatment with the *M. esculenta* extract also prevented the tissue lipid peroxidation.

The mechanism of nephroprotective action of the *M. esculenta* extract, was studied using relative changes in intracellular ROS in the kidney cells monitored using a fluorescent probe, DCFH-DA. DCFH-DA diffuses through the cell membrane readily and is hydrolyzed by intracellular esterases to non fluorescent DCFH, which is then rapidly oxidized to highly fluorescent DCF in the presence of ROS. The DCF fluorescence intensity is proportional to the amount of ROS formed intracellularly. Treatment with *Morchella esculenta* extract was reported to reduce the ROS level to almost the normal when compared to control group of animals which received only cisplatin and gentamicin. A significant reduction in relative fluorescent intensity (RFU) was noticed in the extract treated animals as compared to cisplatin or gentamicin alone treated animals. The result clearly signifies the ability of the extract to scavenge or reduce the production of ROS, resulting from the oxidative stress induced by cisplatin and gentamicin in kidney cells. The extract significantly protects the kidney cells from drug induced oxidative stress leading to kidney damage. The study suggest that aqueous-ethanol extract of morel mushroom, *M. esculenta* mycelium prevent cisplatin and gentamicin induced nephrotoxicity possibly by enhancing renal antioxidant system [24].

Grifola Frondosa

Grifola frondosa, commonly known as the dancing mushroom or Maitake is regarded to impart vitality to health. It was shown that β-glucan from maitake (*Grifola frondosa*) possesses the nephroprotective effects against cisplatin induced nephrotoxicity . A β-glucan purified from *G. frondosa* enhances the efficacy of anti-cancer agent cisplatin, checking the decrease in the number of immunocompetent cells, viz. macrophages, DCs and NK cells in cisplatin-treated mice [25.]

Conclusion

Medicinal mushrooms represent a growing segment of today's pharmaceutical industry owing to the plethora of useful bioactive compounds. While they have a long history of use across diverse cultures, they are backed up by reasonable scientific investigation now. The mycologists around the world firmly believe that a greater knowledge of mushroom can ameliorate many forms of diseases. Exploration of unique species with medicinal properties from the untapped wilderness is warranted. Studies to date have identified a number of compounds and elucidated underlying mechanism. However, research is needed to elucidate the different roles of multiple active compounds from mushrooms and the pathways involved. These findings are important because most of the patients are suffering from drug induced kidney damages and most of the drugs incuding antibiotics, chemotherapeutic agents, NSAIDs used today can induce kidney toxicity. The present results and data might provide new insights into the possible therapeutic uses of mushrooms and helpful suggestions for the design of nehroprotective agents from mushrooms which can ameliorate the side effects of many commonly used drugs.

References

1. Barros, L., Baptista, P., Estevinho, L.M., Ferreira ICFR., Bioactive properties of the medicinal mushroom *Leucopaxillus giganteus* mycelium obtained in the presence of different nitrogen sources. *Food Chem.*, 2007, **105**:179–186.

2. Sarikurkcu, C., Tepe, B., Yamac, M., Evaluation of the antioxidant activity of four edible mushrooms from the Central Anatolia, Eskisehir–Turkey: *Lactarius deterrimus, Suillus collitinus, Boletus edulis,Xerocomus chrysenteron. Bioresour Technol.*, 2008, **99**: 6651–6655.

3. Wang, Z., Luo, D., Liang, Z., Structure of polysaccharides from the fruiting body of *Hericium erinaceus Pers. Carbohydrate Polym.*, 2004, **57**, 241–247.

4. Kim, H.Y., Yoon, D.H., Lee, W.H., Han, S.K., Shrestha, B., Kim, C.H., et al., *Phellinus linteus*inhibits inflammatory mediators by suppressing redox-based NF-jB and MAPKs activation in lipopolysaccharide-induced RAW 264.7 macrophage. *J Ethnopharmacol.*, 2007, **114**, 307–315.

5. Synytsya, A., Mickova, K., Synytsya, A., Jablonsky, I., Spevacek, J., Erban, V., Glucans from fruit bodies of cultivated mushrooms *Pleurotus ostreatus* and *Pleurotus eryngii*: structure and potential prebiotic activity. *Carbohydr Polym.*, 2009, **76**, 548–556.

6. Chen, J., Seviour, R., Medicinal importance of fungal β-(1 ● 3) (1 ● 6)-glucans. *Mycol Res.*, 2007, **111**, 635–652.

7. James, W., Lohra, Gail, R., Willsky, and Margaret A. Acara., Renal Drug Metabolism, *Pharmacol Rev.*, 1998, 50, **1**, 107-140.

8. Singh, N.P., Ganguli,A., Prakash, A., Drug Induced Kidney Disease, *JAPI.*, 2003, **51**, 970-979.

9. Sharma, J.R., In: *Ecology and distribution of Hymenochaetaceae, in Hymenochaetaceae of India.* Sharma J.R., editor. Botanical Survey of India; Culcutta, India., 1995, 9–10.

10. Leelavathy, K.M., Ganesh, P.N., *Polypores in Kerala.* Daya Publishing House; Delhi, India., 2000, 50–60.

11. Ajith, T.A., Jose, N., Janardhanan, K.K., Amelioration of cisplatin induced nephrotoxicity in mice by ethyl acetate extract of a polypore fungus, *Phellinus rimosus.* J. *Exp. Clin. Cancer. Res.*, 2002, 21, **2**, 487–491.

12. Sheena, N., Ajith, T.A., Janardhanan, K.K., Prevention of nephrotoxicity induced by the anticancer drug cisplatin, using Ganoderma lucidum, a medicinal mushroom occurring in South India. *Curr Sci.*, 2003, 85,**4**, 478–482.

13. Sheena, N., Lakshmi, B., and Janardhanan, K.K., Therapeutic potential of Ganoderma lucidum (Fr.) P. Karst, *Nat. Prod. Rad.*, 2005,4, **5**, 382-386.

14. Ying-Hua Shieh et al., Evaluation of the Hepatic and Renal-protective Effects of *Ganoderma lucidum* in Mice *Am. J. Chin. Med.*, 2001, **29**, 501

15. Chang, S.T., Global impact edible and medicinal mushrooms on human welfare in the 21st century: non green evolution. *Int. J. Med. Mushr.*, 1999,1: 1-7.

16. Gunde- Cimmerman, N., Medicinal value of the genus Pleurotus (Fr.) P. Kaest (Agaricales s.l., Basidiomycetes). *Int. J. Med. Mushr.*, 1999, **1**, 69-80.

17. Sasikumar, V., and Sudha, G Menon., Antioxidant activity and nephroprotective effects of Aqueous extract of *pleurotus eous* (berk.) Sacc.: (apk1) pink edible oyster mushroom, *Int. J. Pharma. Bio sci.*, 2011,2, **3**, 92-103.

18. Hajizadeh Moghaddam, M., Javaheri1, S,F., Nabavi, M.R., Mahdavi, S.M., Nabavi1, M.A., Ebrahimzadeh., Protective Role Of *Pleurotus Porrigens* (Angel's Wings) Against gentamicin-induced nephrotoxicty in mice. *Eur. Review. Med. Pharmacol Sci.*, 2010, **14**, 1011-1014.

19. Jayakumar, T., Sakthivel, M., Thomas, P.A., Geraldine. P., *Pleurotus ostreatus*, an oyster mushroom, decreases the oxidative stress induced by carbon tetrachloride in rat kidneys, heart and brain. *Chemico-Biol. Inter.*, 2008, <u>176, 2–3</u>, 108–120.

20. Sirag, H.M., Biochemical and Hematological Studies for the Protective Effect of Oyster Mushroom (*Pleurotus ostreatus*) Against Glycerol-Induced Acute Renal Failure in Rats. *J. Biol Sci.*, 2009, 9,7, 746-752.

21. Nunu alam., Rahul Amin., Asaduzzaman Khan., Ismot Ara., Mi Ja Shim., Min Woong Lee, U., Youn Lee., and Tae Soo Lee., Comparative effects of Oyster Mushrooms on Lipid Profile, liver and kidney function in hypocholesterolemic rats. *Microbiol.*, 2009, 37,**1**, 37-42.

22. Arora, D., Mushrooms Demystified; Ten Speed Press: Berkeley, CA., 1986 pp 784-793.

23. Prasad, P., Chauhan, K., Kandari, L.S., , *Morchella esculenta* (Guchhi): Need for scientific intervention for its cultivation in Central Himalaya. *Curr. Sci.*, 2002, **82**, 1098-1100.

24. Nitha, B., Janardhanan, K.K. Aqueous-ethanolic extract of morel mushroom mycelium Morchella esculenta, protects cisplatin and gentamicin induced nephrotoxicity in mice. *Food Chem Toxicol.*, 2008, 46, **9**, 3193-9.

25. Masuda, Y., Inoue, M., Miyata, A., Mizuno, S., and Nanba, H., Maitake β-glucan enhances therapeutic effect and reduces myelosupression and nephrotoxicity of cisplatin in mice. *Int.J. Immunopharmacol.*, 2009, **9**, 620-626.

11

Influence of substrate and supplementation on mycelial growth and yield of *Lentinus edodes*

Ajay Singh, Sujata Makkar, Anjum Varshney, Jitendra Kumar and
Manjit Singh*
HAIC Agro Research & Development Centre, Murthal, Sonipat (HR) India
* Directorate of Mushroom Research, (ICAR) , Chambaghat, Solan, (HP) India

Shitake mushroom, *Lentinus edodes* (Berk.) was cultivated on sawdust of various fruit and timber trees to see the influence on mycelial growth rate and biological yield. Fastest mycelial colonization, bump formation and early primordia initiation was recorded on Mango (*Mangifera indica*) sawdust supplemented with 5 and 10% wheat bran followed by growth on sawdust of Kail (*Pinus excelsa*) and Meranti (*Shorea spp.*). Highest biological yield was obtained from mango sawdust supplemented by 5% wheat bran followed by Kail and Meranti sawdust supplemented with 5 and 15% wheat bran respectively. Poorest biological yield was obtained from Safeda (*Eucalyptus globosus*) sawdust.

Introduction

The Mushrooms are becoming a regular part of sophisticated households these days as food in the kitchen or biotechnologically produced food products from lignocelluloses. These have become commercially an economically profitable venture all over the world now-a-days. Among the popular varieties of mushrooms– *Lentinus edodes* (Berk.), Shiitake mushroom is the 2nd most important cultivated mushroom after *Agaricus bisporus,* Button Mushroom and one of the most popular edible mushrooms in the world because of its high nutritional value and medicinal properties [1,2]. It grows best in the winter season but can be cultivated all over the year under controlled condition. It is commercially highly viable because it has a shelf life which is more than other varieties of mushroom. Shiitake mushroom has immense health benefits. It enhances anti-cancerous, anti-diabetic, hypocholesterolemic, antimicrobial activities, thereby boosting immune system [3]. Also it is important nutritionally because of its higher protein, dietary fibers and mineral contents [4]. For healthy growth during cultivation of mushrooms need a source of lignocellulose material which supports growth, development and fruiting of mushroom[5] . In this mushroom, sawdust forms the base of the substrate [6-8] . This substrate can be further activated by the use of starch-based supplements such as wheat bran, rice bran, millet, rye or corn. These supplements can be added at different concentration of dry weight to the main ingredient[9-11]. Cultivation has a special relevance to India, because sawdust and other materials are abundantly available to our farmers. Sawdust of commonly available four common fruit & timber trees viz. Mango (*Mangifera indica),* Kail (*Pinus excels)* Meranti (*Shorea spp.),* Safeda (*Eucalyptus globosus)* has been taken to see influence of substrate on the growth & yield of shiitake mushroom.

Experimental

The experiment was carried out at Integrated Mushroom R&D Centre, HAIC Agro Research & Development Centre, Murthal, India. Culture of Shiitake mushroom brought from Directorate of Mushroom Research, ICAR, Solan & sawdust of mango, kail, meranti & safeda were procured from saw mills around Sonipat.

From culture grain spawn was made by following standard process of spawn production.

Four types of Sawdust used as substrates were thoroughly cleared from foreign material & wood parts. Heaps of sawdust of each type was made separately & water equal to wt. of sawdust was added & mixed. Then each type of sawdust was supplemented with wheat bran at 5, 10 & 15% on w/w basis. Moisture & pH of substrate was maintained in the range of 58-60% & 6.5-7.0 respectively. These were left as such overnight & then substrate was filled in autoclavable polypropylene bags with ten replications of each treatment (750gms each). All types of substrate were autoclaved at 22 psi for two hours. After 72 hrs. of autoclaving, substrate bags were inoculated with spawn of shiitake @ 2% at laminar flow hood. Then inoculated bags were incubated at 25°C for colonization of substrate by mycelium of shiitake (Figure 1). At complete colonization, browning & thereafter bump formation, bags were slit open & dipped in ice water for 15 minutes i.e. shock treatment was given on removal of PP bags. Temperature & relative humidity of growing chamber was maintained at 15 ± 2^0C & RH 80-90%. Sufficient water was applied and proper aeration was maintained in culture house to release CO_2 and the supply of O_2 for primordial initiation and fruiting body development. To ensure a homogenous production, it is necessary to give a thermal shock which consists in subjecting the substrate to a sudden change in temperature that is lowering the temperature of substrate to 4-10°C. These bags were then kept at a temperature of 4-10°C for 24 hr. Then the bags taken to fructification room and the plastic bags were taken off. During the fruiting process, temperature of 15 to 20°C, humidity of 80-93% was maintained in almost total darkness. Mushroom fruiting bodies were harvested when the mushroom cap surface was flat to slightly up-rolled at the cap margins. The yield and biological efficiency of mushrooms were recorded regularly.

Time required from inoculation to primordial initiation.

The shortest time for primordial initiation was on the Mango substrate followed by Kail and Meranti as shown in the table 1. The longest time for primordial initiation was on the Safeda.

Fig. 1: Colonization of Shitake mycelium and fruiting body production.

Table 1: Effect of substrate supplementation on the growth and yield of *Lentinus edodes*

Substrate	Supple mentat- ion level of wheat bran*	Time required from inoculation to primordial initiation (days)	Time required from inoculation to first harvest (days)	Number of fruiting bodies per bag	Weight of fruiting bodies per bag (g)	Biolo- gical Efficiency (%)	Biological Yield (%)
Mango	0%	85	103	4	160.73	35.72	21.43
	5%	<u>79</u>	100	13	654.75	145.2	87.3
	10%	84	102	7	306.59	68.13	40.88
	15%	85	103	3	103.21	22.94	13.76
Kail	0%	103	161	2	48.34	10.74	10.45
	5%	90	150	6	274.26	60.95	36.57
	10%	98	155	5	205.33	45.63	27.38
	15%	101	158	2	68.79	15.29	9.17
Safeda	0%	119	-	-	-	-	-
	5%	115	160	3	108.19	24.04	14.42
	10%	116	165	1	16.11	3.58	2.15
	15%	118	165	1	12.67	2.81	1.69
Meranti	0%	103	161	1	15.61	3.47	2.08
	5%	90	150	3	107.20	23.82	14.29
	10%	98	155	4	158.06	35.15	21.07
	15%	101	158	5	223.46	49.66	29.79

*n=10

Biological Efficiency = weight of fresh mushroom/ weight of dry substrate x100

Biological Yield = weight of fresh mushroom/ weight of total substrate x100

Time required from inoculation to first harvest.

The duration from inoculation of spawn packet to first harvest on the substrates of Mango, Kail, Meranti and Safeda are 100, 150, 150 and 160 days respectively. The shortest time to first harvest was on Mango substrate followed by Kail and Meranti and the longest time was taken by Safeda substrate.

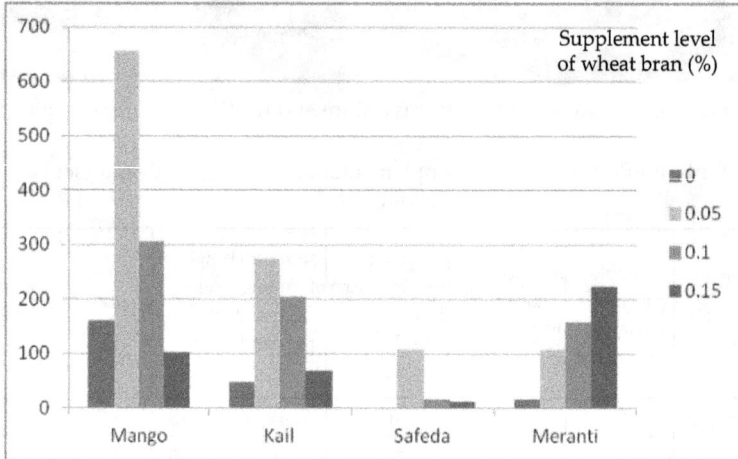

Fig. 2a: Weight of fruiting bodies per bag in g

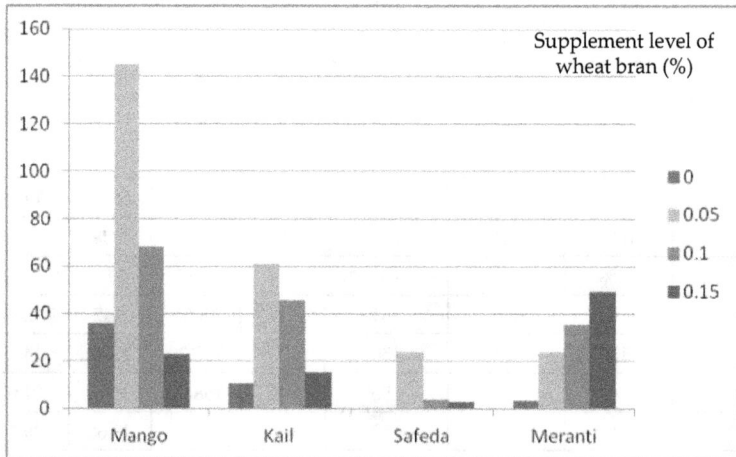

Fig. 2b: Biological efficiency

Number of fruiting bodies per bag

In case of Mango, maximum number of fruiting bodies was formed at 5% WB supplementation level but on Meranti substrate, maximum number of fruiting bodies was formed at 15% WB supplementation level.

Weight of fruiting bodies per bag

In case of Mango, largest quantities of fruiting bodies was formed at 5% WB supplementation level but on Meranti substrate, largest quantities of fruiting bodies was formed at 15% WB supplementation level.

Biological Efficiency

The biological efficiency was found maximum with 5% WB supplementation level in case of Mango, Kail and Safeda while Meranti, showed highest biological efficiency at 15% WB supplementation level.

Biological Yield

Biological yield was found highest with 5% in case of Mango, Kail and Safeda except Meranti, which showed highest at 15 %.

Conclusion

Use of Mango sawdust as a substrate supplement with 5% & 10% WB appears to be most suitable because it gives maximum biological yield and highest level of biological efficiency. It is for this reason that is it is the most suitable for large scale production of shiitake mushroom. Moreover it is highly beneficial because it takes minimum time to fructify. The result of this research leads us to arrive at the conclusion that Mango substrate with 5% WB can be highly useful for large scale production of shiitake mushroom.

References

1. Chang, S.T. and Buswell, J.A., Mushroom nutriceuticals. *World J.Microbiol. Biotechnol.* 1996, **12**, 473–476.

2. Miles, P.G. and Chang, S.T., Genetics of fungi. In: Miles, P.G., Chang, S.T. (Eds.), Mushroom biology: concise basics & current developments, *World Scientific Singapura.* 1997, **194**, 65–82.

3. Wasser, S.P., Shiitake (*Lentinus edodes*). Encyclopedia of Dietary supplements. In: Coates, P., Blackman, M.R., Cragg, G., Levine, M., Moss, J., White, J. (Eds.), Informaworldonline publication [ISBN (electronic) 978-0-8247-5503-4], 2005, pp. 653–664.

4. Khan, M.A., Khan, L.A., Hossain, M.S., Tania, M. and Uddin, M.N., Investigation on the nutritional composition of the common edible and medicinal mushrooms cultivated in Bangladesh. *Bangladesh J. Mushr.* 2009,**3**, 21–28.

5. Chang S.T, and Miles, P.G., Mushrooms: Cultivation, nutritional value, medicinal effect, and environmental impact, 2nd ed. CRC Press, Boca Raton, FL. 2004.

6.   Miller M.W., and Jong S.C., Commercial cultivation of shiitake in sawdust filled plastic bags. Dev-Crop-Sci. Amsterdam: Elsevier Scientific Pub. Co. 1987,**10,**421-426.

7.   Palomo A, Door C, and Mattos L., Comparative study of different substrates for the growth and production of *Lentinus edodes* Berk ("Shiitake"). *Fitopatologia,* 1998,**33,** 71-75.

8.   Grodzinskaya, A.A., Infante, H.D. and Piven, N.M., Cultivation of edible mushrooms using agricultural and industrial wastes. *Agronomia- Tropical-Maracay,* 2003,**52,** 427-447.

9.   Ivan, H.R., Antonio, C.M., Jose, O.M. and Jose, C.B., Supplementation of sugarcane bagasse with rice bran and sugarcane molasses for shiitake (*Lentinula edudes*) spawn production. *Brazil J. Microbiol.*2003, **34**: 61-65.

10. Royse, D.J., Bahler, B.D. and Bahler, C.C., Enhanced yield of shiitake by saccharide amendment of the synthetic substrate. *Appl. Environ. Microbiol.* 1990,**56,** 479-482.

11. Royse, D.J., Yield stimulation of shiitake by millet supplementation of wood chip substrate. *Mushr Biol. Mushr. Prod.*1996, **2**: 277-283.

12

Ganoderma lucidum triterpenes – New scope in cancer therapy

T.P. Smina[1] and K.K. Janardhanan[2*]

[1]CeNTAB, SASTRA University, Thanjavur, Tamilnadu, India.
[2]Amala Cancer Research Centre, Amala Nagar, Thrissur, Kerala, India, 680 555.
Email: kkjanardhanan@yahoo.com.

For millennia, mushrooms and their metabolites are used for the treatment of a variety of human ailments. Experimental studies had addressed the biological activities of whole mushrooms or crude extracts by oral administration to humans. *Ganoderma lucidum* or Reishi, also known as 'King of herbs', is a polypore mushroom, well known for its adaptogenic properties. *G. lucidum* products have been widely used as a single agent or in combination with other herbal medicines or chemotherapeutic drugs for many years, mainly in Asian countries. It is an alternative adjuvant in the treatment of leukaemia and carcinoma. In this review, we summarize the main biological activity of triterpenes isolated from *G. lucidum* with a special emphasis on its anticancer activity.

Introduction

Mushrooms contain a large number of active compounds and hence are considered as a promising resource of physiologically functional food and as material for the development of medicines and dietary supplements. For millennia, mushrooms and their metabolites are used for the treatment of a variety of human ailments. However, there are few experiments about epidemiologic and experimental studies that address the biological activities of whole mushrooms or crude extracts by oral administration to humans.

Ganoderma lucidum or Reishi, also known as 'King of herbs',is a polypore mushroom, well known for its adaptogenic properties. This important medicinal mushroom is variously reported to have multi-beneficial effects on viscera, audio, visual and olfactory senses, to improve intelligence, enhance the memory and to retard aging process. In clinical studies, *G. lucidum* products have been widely used as a single agent or in combination with other herbal medicines or chemotherapeutic drugs for many years, mainly in Asian countries. *G. lucidum* has been used in folk medicine of China and Japan, especially in the treatment of hepatopathy chronic hepatits, nephritis, hypertension, asthma etc.[1,2]. Medicinal uses of *G. lucidum* in ancient Far East countries included the treatment of neurasthenia, anorexia, mushroom poisoning (antidote), coronary heart diseases, carcinoma and bronchial cough etc.[2,3,4]. *G. lucidum* has been reported to have a number of pharmacological effects including immunomodulating, anti-inflammatory, analgesic, chemopreventive, antitumour, radioprotective, sleep promoting, antibacterial, antiviral (including anti-HIV), hypolipidemic, antifibrotic, hepatoprotective, anti-oxidative/ radical scavenging with anti-ageing, hypoglycemic and anti-ulcer properties. Reishi has now become recognized as an alternative adjuvant in the treatment of leukemia, carcinoma, hepatitis and diabetes[1-9]. In this review we try to summarize the main biological activity of triterpenes isolated from *G. lucidum* with a special emphasis on its anticancer activity.

Medical mushroom produce a variety of biologically active compounds such as polysaccharides, lectins, terpenoids, phenols etc. Over 300 reports have been published concerning the chemical constituents of *G. lucidum* and related species. There are qualitative and quantitative differences in the chemical composition of *G. lucidum* products depending on the strain, origin, extracting process and cultivation conditions[3,5,10,11]. The fruiting body, mycelia and spores of *G. lucidum* contain approximately 400 different bioactive compounds which mainly include triterpeniods, polysaccharides, nucleotides, sterols, fatty acids, protein/peptides and trace elements[7,8,10,12]. Polysaccharides and triterpenes are two major categories of the bioactive ingredients. The extraction process is necessary to purify the mushroom by eliminating unnecessary components that are produced by the mushroom's natural growing process, while preserving the essential bioactive components within it. Numerous means of extraction have been developed with the objective of obtaining extracts with better yields and lower costs[13]. Extractions using ethanol have been reported to be the easiest approach to maintain the activity of the triterpene extracts and to scale up its production[14].Triterpenes are usually extracted using organic solvents such as methanol, ethanol, chloroform or ether followed by different separation methods[15]. Ultrasonic techniques are currently being used to enhance the rate of extraction of triterpenes by destroying the dense structure in the cells[16].

At least 140 different triterpenes have been identified in *G. lucidum*[2,3,5,10,12]. The majority is bitter tasting and largely occur as ganoderic[12]. Terpenoids are comprised of four groups: the (a) volatile mono- and sesquiterpenes (essential oils) (C_{10} and C_{15}), (b) less volatile diterpenes (C_{20}), (c) in volatile triterpenoids and sterols (C_{30}) and (d) the carotenoid pigments (C_{40}). Most investigations on *Ganoderma* concern the less volatile triterpenoid (triterpene) and sterol forms[17]. Min et al.[18,19]reported the isolation of seven new lanostane – type triterpenes from the fruiting body and also from the spores (ganoderic acids χ, δ, ε, ϕ, γ, η and φ). The spores also contain triterpene lactones[12]. Triterpenoids from *Ganoderma lucidum* have been divided into 10 groups based on the structural similarities and known biological and medicinal properties (Fig.1)[20]. Preliminary studies indicate that the spores contain considerably higher contents of ganoderic acids than other parts of the fungus and that triterpene composition of the fruit body varies according to the area in which it is grown[18].

Therapeutic applications of terpenoid compounds

Anticancer and cytotoxic activities

Triterpenes are primarily isolated from the spores of *G. lucidum* and have shown remarkable pharmacological and therapeutic activities on multiple human

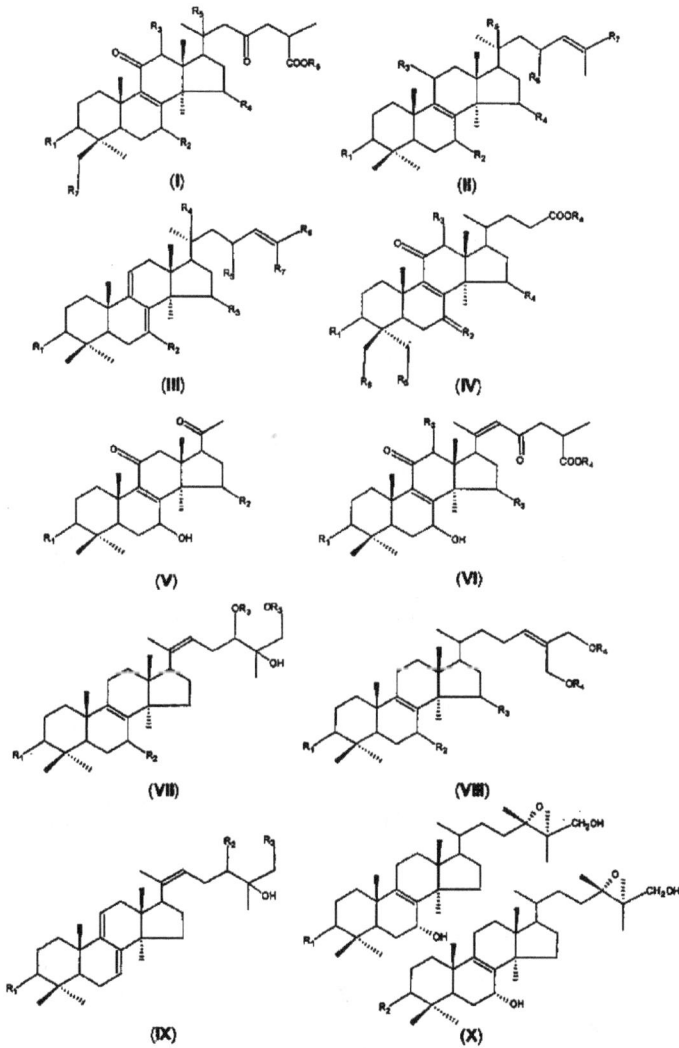

Fig. 1: Ten groups of lanostane-type triterpenoids of
Ganoderma lucidum (Wasser, 2005)

diseases including cancer[21,22]. Studies have shown that many subtypes of triterpene extracts from *G. lucidum* can directly induce apoptosis of multiple human cancer cell lines[21].*Ganoderma lucidum* triterpenes suppressed cell proliferation of RAW264.7 cells through cell cycle arrest at G0/G1–G2M, which was mediated by the down-regulation of expression of cell cycle regulatory proteins cyclin D1, CDK4 and cyclin B1, respectively[23]. In another study, *G. lucidum* triterpenoid extract induces apoptosis in human colon carcinoma cells (Caco-2)[24]. The triterpene enriched fraction, WEES-G6, was prepared from mycelia of *G. lucidum* by sequential hot water extraction, removal of ethanol-insoluble polysaccharides and then gel-filtration chromatography found to inhibit the growth of human hepatoma Huh-7 cells, but not Chang liver cells, a normal human liver cell line[25].

Ganoderic acid A modulate AP-1 and NF-kB signaling, leading to suppression of growth and invasive behaviour in cancer[25-28]. Five purified ganoderic acids, ganoderic acid F (GAF), ganoderic acid K (GAK), ganoderic B (GAB), ganoderic acid D (GAD) and ganoderic acid AM1 (GAAM1) treated for 48 h inhibited the proliferation of HeLa human cervical carcinoma cells by altering proteins involved in cell proliferation and/or cell death, carcinogenesis, oxidative stress, calcium signalling and ER stress[29]. Ganoderic acid D has been shown to directly bind to 14-3-3ζ protein and this binding may contribute to the apoptosis observed in HeLa cell[30]. Ganoderiol F (GA-F) is a tetracyclic triterpene found in *Ganoderma* species[19,31]that showed cytotoxicity *in vitro* against Lewis lung carcinoma (LLC), Meth-A, Sarcoma-180, and T-47D cell lines[8,19,32]. In addition, the anti-tumour effect of GA-F has been demonstrated *in vivo* in rats implanted with LLC tumour cells[33].

Ganoderic acid Me (GA-Me) is a lanostane triterpenoid purified from *G. lucidum* mycelia inhibited multidrug resistance and induced apoptosis in multidrug resistant colon cancer cells[34]. In another study by Chen and Zhong[35], Ganoderic acid Me (GA-Me) induced p53-mediated sub-G1 arrest in human colon cells. Ganoderic acid Mf and S also induced mitochondria mediated apoptosis in human cervical carcinoma HeLa cells[36]. Ganoderic acid T is the most abundant triterpenic acid found in *G. lucidum* and shows significant anti-cancer effects in both *in vitro* and *in vivo* studies[37,38]. It also showed anti-invasion and antimetastasis *in vitro*[39]. Triterpenoids, such as ganoderic acids T–Z isolated from *G. lucidum*, showed cytotoxic activity *in vitro* on hepatoma cells[40].

Lucidenic acid N, lucidenic acid A, and ganoderic acid E, three triterpenoids, isolated from the dried fruiting bodies of *G. lucidum* showed significant cytotoxic activity against Hep G2 and P-388 tumorcells[41]. Lucialdehydes B, C, ganodermanonol and ganodermanondiol isolated from the fruiting bodies of *G. lucidum* showed cytotoxic effects *in vitro* against Lewis lung carcinoma (LLC), T-47D, Sarcoma 180, and Meth-A tumor cell lines. Of the compounds, lucialdehyde C exhibited the most potent cytotoxicity against LLC, T-47D, Sarcoma 180, and Meth-A tumor cells with ED_{50} values of 10.7, 4.7, 7.1, and 3.8m g/ml, respectively[8,32]. A lanostanoid, 3b-hydroxyl-26-oxo-5a-lanosta-8,24-dien- 11-one, and a steroid, ergosta-7,22-diene-3b,3a,9atriol, isolated from fruiting bodies of *G. lucidum*, demonstrated potent inhibitory effects on KB cells and human PLC=PRF=5 cells *in vitro*[42]. Additional forms of triterpenes isolated from *G. lucidum* have shown cytotoxicity in the following human cancer cell lines: p388, HeLa, BEL-7402, and SGC-7901[43,44].

Other therapeutic activities

G. lcidum triterpenes markedly suppressed the secretion of inflammatory cytokine tumor necrosis factor-α (TNF-α) and interleukin-6 (IL-6), and inflammatory mediator nitric oxide (NO) and prostaglandin E2 (PGE2)from lipopolysaccharide (LPS)-stimulated murine RAW264.7 cells. Apart from its anti-inflammatory activity, *G. lucidum* triterpenes suppressed cell proliferationof RAW264.7 cells[23].

Several triterpenes from *G. lucidum* (M. A. Curtis:Fr.) P. Karst. [i.e. ganoderiol F, ganodermanontriol, ganoderic acid B] are active as antiviral agents against human immunodeficiency virus type 1 (HIV-1). The minimum concentration of ganoderiol F and ganoderma nontriol for complete inhibition of HIV-1 induced cytopathic effect in MT-4 cells is 7.8 mg/ml. Ganoderic acid B inhibits HIV-1 protease with an IC 50 value of 0.17 mM[45].

Ganoderic acids R and S and ganosporeric acid-A from *G. lucidum* show *in vitro* anti hepatotoxic activity in the galactosamine-induced cytotoxic test with primary cultured rat hepatocytes[46,47]. *In vivo* two fractions of a total triterpenoids extract of *G. lucidum* (75% ethanol) can protectmice against hepatic necrosis induced by chloroform and D-galactosamine.

References

1. Jong, S.C., Birmingham J.M. Medicinal benefits of the mushroom *Ganoderma*. *Adv. Appl. Microbial.*, 1992,**37**, 101-134.

2. Chang, S.T. Global impact of edible and medicinal mushrooms on human welfare in the 21st century: Non green revolution. *Int. J. Med. Mushrooms.*, 1999,**1**,1–7.

3. Wasser, S. P., Weis, A. L. Medicinal mushrooms. Reishi mushroom (*Ganodermalucidum* (Curtis: Fr.) P.Karst), Nevo E, Ed., Peledfus, Haifa. 1997.

4. Zhou, S. H., Kestell, P., Baguley, B. C., Paxton, J. W. 5,6-Dimethylxanthenone-4-acetic acid: a novel biological response modifier for cancer therapy. *Invest. New Drugs.*, 2002, **20**, 281–295.

5. Hobbs, C. H. Medicinal mushrooms: An exploration of tradition, healing and culture. Botanica Press, Santa Cruz,CA. 1995, pp 251.

6. Wasser, S. P., Weis, A. L. Therapeutic effects of substances occurring in higher basidiomycetes mushrooms: a modern perspective. *Crit. Rev. Immunol.*, 1999,**19**, 65–96.

7. Smith, J. E., Rowan, N. J., Sullivan, R. Medicinal mushrooms: a rapidly developing area of biotechnology for cancer therapy and other bioactivities. *Biotechnol. Lett.*, 2002,**24**, 1839-1845.

8. Gao, J. J., Min, B. S., Ahn, E. M., Nakamura, N., Lee, H. K., Hattori, M. New triterpene aldehydes, lucialdehydes A-C, from *Ganodermalucidum* and their cytotoxicity against murine and human tumor cells. *Chem. Pharm. Bull.*, 2002,**50**, 837-840.

9. Gao, Y., Zhou, S., Jiang, W., Huang, M., Dai, X. Effects of ganopoly (a *Ganodermalucidum* polysaccharide extract) on the immune functions in advanced-stage cancer patients. *Immunol. Invest.*, 2003,**32**, 201–215.

10. Mizuno, T., Sakai, T., Chihara, G. Health Foods and medicinal usages of mushrooms. *Food. Rev. Int.*, 1995,**11**, 69-81.

11. Stamets, P. Growing Gourmet and Medicinal Mushrooms, 3rd Ed.; Ten Speed Press: CA, USA. 2000.

12. Kim, H. W., Kim, B. K. Biomedical triterpenoids of *Ganodermalucidum* (Curt.: Fr.) P. Karst. (Aphyllophoromycetideae). *Int. J. Med. Mush.*, 1999,**1**,121-138.

13. RuhanAskin, M. S., Goto, M. Extraction of bioactive compounds from *Ganodermalucidum*. Department of Applied Chemistry and Biochemistry, Kumamoto University. 2008.

14. Gao, Y., Zhang, R., Zhang, J., Gao, S., Gao, W., Zhang, H., Wang, H., Han, B. Study of the extraction process and in vivo inhibitory effect of *Ganoderma*triterpenes in oral mucosa cancer. *Molecules.*, 2011, **16**,5315-5332.

15. Gao, Y., Zhou, S. Cancer Prevention and Treatment by *Ganoderma*, a Mushroom with Medicinal Properties. *Food Reviews International.*, 2003, **19**,275-325.

16. Zhou, X. W., Su, K.Q., Zhang, Y. M. Applied modern biotechnology for cultivation of *Ganoderma* and development of their products. *Applied Microbiology and Biotechnology.*, 2012, **93**,941-963.

17. Lindequist, U. Structure and biological activity of triterpenes, polysaccharides and other constituents of *Ganodermalucidum*. In: Kim, B. K., Kim, I. H., Kim, Y. S. (Eds.), Proceedings of the 6th International Symposium on Recent Advances in *Ganodermalucidum* Research. The Pharmaceutical Society of Korea, Seoul, 1995, pp. 61– 91.

18. Min, B. S., Nakamura, N., Miyashiro, H., Bae, K. W., Hattori, M. Triterpens from the spores of *Ganodermalucidum* and their inhibitory activity against HIV-1 protease. *Chem. Pharm. Bull.*, 1999,**46**, 1607–1612.

19. Min, B. S., Gao, J. J., Nakamura, N., Hattori, M. Triterpenes from the spores of *Ganodermalucidum* and their cytotoxicity against meth-A and LLC tumor cells. *Chem. Pharm. Bull.*, 2000, **48**, 1026-1033.

20. Wasser, S. P. Reishi or Ling Zhi (*Ganodermalucidum*). In: Coates, P. M., Blackman, M. R., Cragg, G. M., Levine, M., Moss, J., White, J. D. (Eds.) Encyclopedia of Dietary Supplements. Marcel Dekker, NY, USA, 2005. pp. 603–622.

21. Yuen, J. W., Gohel, M. D. Anticancer effects of *Ganodermalucidum*: a review of scientific evidence. *Nutr Cancer.*, 2005,**53**, 11-17.

22. Shi, L., Ren, A., Mu, D., Zhao, M. Current progress in the study on biosynthesis and regulation of ganoderic acids. *Appl. Microbiol. Biotechnol.*,2010,**88**,1243-1251.

23. Dudhgaonkar, S., Thyagarajan, A., Sliva, D. Suppression of the inflammatory response by triterpenes isolated from the mushroom *Ganodermalucidum*. *Int. ImmunoPharmacol.*, 2009,**9**, 1272–1280.

24. Ruan, W., Popovich, D. G. *Ganodermalucidum*triterpenoid extract induces apoptosis in human colon Carcinoma cells (Caco-2). *Biomedicine & Preventive Nutrition.*, 2012,**9**, 203–209.

25. Shiao, M. S. Natural products of the medicinal fungus *Ganodermalucidum*: Occurrence, biological activities, and pharmacological functions. *Chem. Rec.*, 2003,**3**, 172-180.

26. Sliva, D. *Ganodermalucidum* (Reishi) in cáncer treatment. *Integr. Cancer. Ther.*, 2003,**2**, 358-364.

27. Jiang, J., Grieb, B., Thyagarajan, A., Sliva, D. Ganoderic acids suppress growth and invasive behavior of breast cancer cells by modulating AP-1 and NF-kBsignaling. *Int. J. Mol. Med.*, 2008,**21**, 577-584.

28. Weng, C. J., Chau, C. F., Hsieh, Y. S., Yang, S. F., Yen, G. C. Lucidenic acid inhibits PMA-induced invasion of human hepatoma cells through inactivating MAPK/ERK signal transduction pathway and reducing binding activities of NF-kB and AP-1. *Carcinogenesis.*, 2008,**29**, 147-156.

29. Yue, Q. X., Song, X. Y., Maa, C., Feng, L. X., Guan, S. H., Wu, W. Y., Yang, M., Jiang, B. H., Liu, X., Cui, Y. J., Guo, D. A. Effects of triterpenes from *Ganodermalucidum* on protein expression profile of HeLacells. *Phytomed.*, 2010, **17,** 606–613.

30. Yue, Q. X., Cao, Z. W., Guan, S. H., Liu, X. H., Tao, L., Wu, W. Y., Li, Y. X., Yang, P. Y., Liu, X., Guo, D. A. Proteomics characterization of the cytotoxicity mechanism of ganoderic acid D and computer-automated estimation of the possible drug target network. *Mol. Cell. Proteomics.*, 2008,**7**, 949-961.

31. Chang, U. M., Li, C. H., Lin, L. I., Huang, C. P., Kan, L. S., Lin, S. B.Ganoderiol F, a *Ganoderma*triterpene, induces senescence in hepatoma HepG2 cells. *Life Sci.*, 2006,**79**, 1129-1139.

32. Gao, Y., Zhou, S., Chen, G., Dai, X., Ye, J. A phase I/II study of a *Ganodermalucidum*(Curt.: Fr.) P. Karst. extract (Ganopoly) in patients with advanced cancer.*Int. J. Med. Mushrooms.*,2002,**4,**207–214.

33. Gao, J. J., Hirakawa, A., Min, B. S., Nakamura, N., Hattori, M. *In vivo* antitumor effects of bitter principles from the antlered form of fruiting bodies of *Ganodermalucidum. J. Nat. Med.*, 2006,**60**, 42-48.

34. Jiang, Z., Jin, T., Gao, F., Liu, J., Zhong, J., Zhao, H. Effects of Ganoderic acid Me on inhibiting multidrug resistance and inducing apoptosis in multidrug resistant colon cancer cells. *Process Biochem.*, 2011,**46**, 1307–1314.

35. Chen, N. H., Zhong, J. J. Ganoderic acid Me induces G1 arrest in wild-type p53 human tumor cells while G1/S transition arrest in p53-null cells. *Process Biochem.*, 2009,**44**, 928–33.

36. Liu, R. M., Zhong, J. J. Ganoderic acid Mf and S induce mitochondria mediated apoptosis in human cervical carcinoma HeLa cells. *Phytomedicine.*, 2011,**18,** 349–355.

37. Chen, N. H., Liu, J. W., Zhong, J. J. Ganoderic acid T inhibits tumor invasion in vitro and *in vivo* through inhibition of MMP expression. *Pharmacol. Rep.*, 2010,**62**, 150-163.

38. Tang, W., Liu, J. W., Zhao, W. M., Wei, D. Z., Zhong, J. J.Ganoderic acid T from *Ganodermalucidum* mycelia induces mitochondria mediated apoptosis in lung cancer cells. *Life Sci.*, 2006,**80**, 205-211.

39. Xu, K., Liang, X., Gao, F., Zhong, J., Liu, J.Antimetastatic effect of ganoderic acid T in vitro through inhibition of cancer cell invasion. *Process Biochemistry.*, 2010,**45**, 1261-1267.

40. Toth, J. O., Luu, B., Ourisson, G. Ganoderic acid T and Z: cytotoxic triterpenes from *Ganodermalucidum* (Polyporaceae). *Tetrahedron Lett.*, 1983, **24**, 1081–1084.

41. Wu, T. S., Shi, L. S., Kuo, S. C. Cytotoxicity of *Ganodermalucidum*triterpenes. *J. Nat. Prod.*, 2001,**64**, 1121-1122.

42. Lin, C. N., Tome, W. P., Won, S. J. Novel cytotoxic principles of Formosan *Ganodermalucidum*. *J. Nat. Prod.*, 1991,**54**, 998–1002.

43. Guan, S. H., Xia, J. M., Yang, M., Wang, X. M., Liu, X., Guo, D. A. Cytotoxic lanostanoidtriterpenes from *Ganodermalucidum*. *J. Asian. Nat. Prod. Res.*, 2008,**10**, 705-710.

44. Niu, X. M., Li, S. H., Xiao, W. L., Sun, H. D., Che, C. T.Two new lanostanoids from *Ganodermaresinaceum*. *J. Asian. Nat. Prod. Res.*,2007, **9**, 659-664.

45. El-Mekkawy, S., Meselhy, M. R., Nakamura, N., Tezuka, Y., Hattori, M., Kakiuchi, N. Anti-HIV-1 and anti-HIV-1-protease substances from *Ganodermalucidum*. *Phytochemistry.*, 1998,**49**, 1651–7.

46. Hirotani, M., Ino, C., Furuya, T., Shiro, M. Ganoderic acids T, S and R, new triterpenoids from the cultured mycelia of *Ganodermalucidum*. *Chem. Pharm. Bull.*,1986,**34**, 2282–2285.

47. Chen, R. Y., Yu, D. Q. Studies on the triterpenoid constituents of the spores from *Ganodermalucidum* Karst. *J. Chin. Pharm. Sci.*, 1993,**2**, 91–96.

13

Cultivation of edible and medicinal mushrooms in Kerala

A. V. Mathew

Department of Plant Pathology, Kerala Agricultural University, Regional Agricultural Research Station, Kumarakom-686 653, Kerala, India

The Sub-tropical climatic conditions prevailing in Kerala is ideal for growing different types of mushrooms. Both terrestrial mushrooms and those growing on wooden logs were collected and brought into pure culture on different tissue culture media. A low cost medium, Potato Sugar Agar enriched with Coconut Water (P.S.A.C.W.) was standardized for culturing different species of mushrooms. The mushrooms isolated were *Pleurotus eous, P.sapidus, Calocybe indica, Tricholoma georgii* and *Ganoderma lucidum*. Their spawn was produced on paddy grains following standard procedures. Cultivation technologies of all the five mushroom species were also standardized. All the species except *G. lucidum* are edible. *G. lucidum* is medicinal and non edible in nature. *P. eous* and *P. sapidus* were grown successfully on paddy straw and rubber wood sawdust. *C. indica* and *T. georgii* were cultivated successfully on paddy straw and banana leaf sheath. The new isolate of *C. indica* could be cultivated throughout the year in the plains, where as the cultivation of *T. georgii* was successful only during the rainy season. Indoor cultivation technology of *T. georgii*, an *ectomycorrhizal* mushroom, was standardized for the first time. *G. lucidum* was cultivated on rubber wood saw dust, enriched with rice bran and a maximum bio-efficiency of 20% was obtained during the rainy season.

Introduction

Mushroom cultivation helps to provide nutritional security and solution for proper recycling of agricultural wastes/byproducts. Mushrooms a highly priced delicacy for more than 3000 years, are now consumed by many people in India. It has high nutritive and medicinal value and contributes to a healthy diet because of its rich source of vitamins, minerals and proteins.[1] *Agaricus bisporus* is the most popular mushroom commercially cultivated all over the world until 1960's. Till now more than 80 edible mushrooms are considered for commercial cultivation. *A. bisporus,*

Pleurotus florida, Calocybe indica and *Volvariella volvacae* are the four major species commercially cultivated in India. The sub-tropical climatic conditions prevailing in Kerala are ideal for growing these species except *A. bisporus*. A survey was carried out in 2009-10 to collect and identify certain local strains/species of edible and medicinal mushrooms so that a calendar could be developed for cultivating mushrooms throughout the year in Kerala. Five different species of mushrooms were collected, brought into pure culture, spawn prepared and domesticated. They are *P. eous, P. sapidus, G. lucidum* (wood mushrooms), *C. indica* and *T. georgii* (terrestrial mushrooms).

Experimental

1. Collection of mushrooms

Surveys were conducted during 2009-10 in the rainy season. Young developing buttons were collected, cleaned, packed properly and brought to the laboratory for isolation and pure culturing.

2. Tissue culture media

Successful mushroom growing depends on the production and maintenance of pure culture and quality spawn. The isolation, purification and maintenance of mushroom cultures require technical expertise and aseptic laboratory facilities. The following culture media were used for tissue culture and spore culture of the collected mushroom species.

   a) P.D.A (Potato-200 g, Dextrose – 20 g, Agar-20 g, D.W. – 1 lit.)

   b) P.S.A (Potato 200 g, Sugar-20 g, Agar-20 g, D.W. 1 lit.)

   c) P.S.A.C.W (Potato 200 g, Sugar-20 g, Agar-20 g, D.W-500ml, Coconut water 500 ml)

   d) O.M.A (Oats meal flakes -30 g, Agar-20 g, D.W- 1lit.)

   e) M.E.A (Malt extract – 20 g, Agar-20 g, D.W-1 lit.)

   P.S.A and P.S.A.C.W are new media whereas the others are existing standard media. These media were prepared as per standard procedures.

3. Culture isolation

a) Vegetative mycelium culture (tissue culture)

All the mushrooms except *G. lucidum* were tissue cultured. Young basidiocarps were collected and the surface cleaned by gently swabbing with cotton dipped in absolute alcohol. Then each species was cut into small pieces of 0.5 cm diameter at the juncture of the stem and the cap. The mushroom pieces were dipped in 0.1% mercuric chloride solution taken in sterile petriplates for 30-60 sec. After the period the mushroom discs were taken out from the mercuric chloride solution and washed in three changes of sterile distilled water (1 min each) taken in sterile Petri plates. The surface sterilized mushroom discs were then plated in different media taken in oven sterilized Petri plates. All these operations were carried out under aseptic conditions in a laminar air flow chamber. The inoculated plates were then incubated at $25 \pm 2°$ C in a BOD incubator. Fresh growth of mycelia observed from these tissues after 4-5 days were transferred and pure cultured. Pure cultures were made by carefully transferring young mycelium from growing edges of the colony from petri plates to test tube slants and again incubated at $25 \pm 2°$ C for 10-14 days.

b) Spore culturing

In the case of *G. lucidum* multispore culturing was made from a fresh spore print/ basidiocarp and suspended in 100ml sterile distilled water in flasks and shaken to obtain uniform spore suspension. A few drops of these suspensions were added to lukewarm culture media taken in oven sterilized petri plates. The plates were rotated to homogenize the spore suspension into the culture media. The culture media were allowed to solidify and incubated at $25 \pm 2°$ C for 3-4 days. The spore germination was observed under a microscope and germinating spores were carefully transferred to test tube slants along with a little medium. The test tube slants were incubated at $25 \pm 2°$C for 10-14 days and further used for producing the mother spawn.

4. Spawn preparation

The seed required for growing mushroom is called the spawn. Usually spawn is prepared by using any cereal grain as the growing medium. In the present study paddy seeds were used for making the spawn of all the five species of mushrooms. Paddy seeds were soaked in water for 8-12 h and after that washed well with freshwater again and again. The seeds were then half cooked by steaming/boiling so that the outer husk is just split open. Then the grains were allowed to cool and drain excess moisture, mixed with 4% calcium carbonate and 1% gypsum (Calcium sulphate). This medium was filled in polypropylene covers (12x6" size) @ 300g/ pkt. and sterilized in an autoclave at $121°$ C for 2h. After cooling the packets, mushroom culture was added to the medium in a laminar air flow chamber on the next day and incubated at room temperature. (25- 28 °C) for 15-20 days. The spawn so prepared is called the Mother spawn. Commercial spawn was prepared from the mother spawn serially transferring up to one or two generations.

5. Cultivation technologies

Six different substrates viz., paddy straw, banana leaf sheath, rubber wood sawdust, coconut sawdust, coir pith and sawdust of Portia tree (*Thespesia populnea*) were used as media for cultivating P. *eous, P. sapidus, C. india and T. georgii.* Paddy straw was soaked for 12 h and banana sheath for 24h in water. Next day both the media were steam sterilized for 45 min, allowed to cool and drain water up to 50% moisture level. The poly bag method of cultivation was used for the studies. Poly bags of 24x12" size was used for the cultivation of *pleurotus spp.* One kilogram dry substrate was used for filling the bags and 150g spawn used for cultivation. For cultivating *C. indica* and *T. georii* poly bags of size 16x13" were used. 750g substrate (dry wt) was used for filling the bags and 150 g spawn was used for cultivating each bag. The *Pleurotus* beds were incubated at a temperature of 26-32 °C and relative humidity of (R.H.) 80-90% in the dark room. Milky mushroom beds were incubated at 28-32°C and 80% R.H. for 20 days. *Tricholoma* beds were incubated at 24-28°C and 90% R.H. for 28 days for spawn running.

6. Casing

Casing means covering the top surface of the beds with pasteurized casing materials after the completion of spawn running. Casing was done for milky mushroom and *Tricholoma* cultivated on paddy straw. The thickness of the casing media applied at the top was 2.5 – 3cm. The following casing media were used in the study i) Soil + sand (3:1) ii) coir pith + soil (1:1) iii) FYM + sandy soil (1:1), iv) coir pith + vermicompost (1:1) and v) coir pith alone. The casing media were mixed with 4%Calcium carbonate and 1% Gypsum (Calcium sulphate) to get a PH of 7.5 – 8 . The media were steam sterilized for an hour and after cooling they were applied at the top of the beds. The casing medium was kept just moist by gently spraying with water every day. The milky mushroom beds were then removed to the cropping room where there was more light (1000 lux), R.H-80-90% and temperature 28-34 °C. *Tricholoma* beds were kept where the temperature was 24-28°C, R.H 90% and light intensity 500 lux.

7. Cultivation of *G. lucidum*

Four different substrates, viz., Rubber wood sawdust, Portia tree saw dust, Coconut sawdust and Coir pith were used as the basic substrates for growing *G. lucidum.* These substrates were separately mixed with rice bran @ 20%, Calcium carbonate 2% and Gypsum 1% to get a pH of 5.5-6 and wetted with water to get a moisture level of 45-50%. The media were separately filled in polypropylene bags (12x6"sze) @2.5kg wet weight/bag (dry wet 1kg.) and their mouths were plugged with non-absorbent cotton and tied with rubber band. The bags were then sterilized in an autoclave at 121°C for 2 h. On the next day, after cooling, the bags were inoculated with mother Spawn @5% on dry wet basis under aseptic conditions in a laminar air flow chamber. The beds were then incubated at 25-30° C in closed rooms with high concentration of Carbon dioxide and darkness. After complete spawn run (20-25 days), the top and bottom portion of the beds were cut at the level of the substrate totally exposing the substrate at the top and bottom and the beds were kept flat on shelves for pinning/fruiting. A temperature of 28-32°C, CO_2 1500 ppm, light 800 lux, and R.H. 90-95% were maintained in the cropping room.

Once, the pinheads have grown up enough to form the cap, indicated by the flattening of the white creamy top of the pinhead, the R.H was reduced to 80% and more fresh air was introduced to get about 1000 p pm Co_2. Once, the cap was fully developed and flattened, indicated by yellowing of the cap margin, temperature and RH were lowered to 25°C, and 65%, respectively for thickening, reddening and maturation of the fruit bodies.

Survey and collection of mushrooms

The following species viz., *P. eous, P. sapidus, C. indica, T. georgii* and *G. lucidum* were collected and identified following morphological characters and comparing with the already available cultures.

Tissue culture

All the species except *G. lucidum* were tissue cultured on all the five media tried. *G. lucidum* was successfully spore cultured on these media. Two new and cheap media tried viz., P.S.A. and P.S.A.C.W. were very successful in supporting the growth of mycelia of all the five species. The best media for mycelia growth of these species were found to be M.E.A and P.S.A.C.W based on the time taken for complete growth of the mycelia in test tube slants.

Spawn Preparation

Mother spawn and commercial spawn were prepared successfully on paddy seeds. Spawn of *P. eous* and *P. sapidus* matured within 10-15 days of inoculation. *Calocybe* and *Tricholoma* spawn took 25-30 days for maturation whereas *Ganoderma* spawn was ready for cultivation within 18-22 days of inoculation.

Cultivation of mushrooms on agro-waste substrates

P. eous and P. sapidus were successfully cultivated on all the six substrates tried. *P. eous* is a short duration variety and it gave first harvesting on paddy straw 10-12 days after spawning (plate 1). On banana sheath first yield was obtained 25-30 days after cultivation. First harvest on rubber wood sawdust was made 18-20 days after spawning (Plate 2). *P.eous and P.sapidus* were also successfully grown in plastic bottles and jars (Plate 3, 4) both on paddy straw and rubber wood sawdust. In the case of *P.sapidus* first harvesting was made 14-16 days after spawning on paddy straw (Plate 5). On banana sheath first harvest was made 22-26 days and on rubber wood sawdust 20-22 days after cultivation (Table 1). Both species were successfully grown throughout the year.

Table 1: Yield and Cropping cycle of *P.eous and P.sapidus* on different substrates

Sl. No.	Substrates	Days to first harvest		B.E.%		Cropping cycle (days)	
		P.eous	*P.sapidus*	*P.eous*	*P.sapidus*	*P.eous*	*P.sapidus*
1	Paddy straw	10-12	14-16	50	40	40-45	40-50
2	Banana sheath	25-30	22-26	40	36	50-60	60-70
3	Rubber wood saw dust	18-20	20-22	80	60	100-120	90-110
4	Coir pith	32-35	30-34	18	19	60-70	65-75
5	Coconut saw dust	30-36	30-36	15	18	60-70	65-75
6	Portia tree saw dust	32-36	28-32	12	10	80-90	80-90

C. indica, the Kumarakom isolate was successfully grown on paddy straw and banana sheath (Plate 6). The other substrates were not economical for its growth and yield. It was cultivated successfully in summer and rainy season. (Table2). All the casing media used for cultivating *C. india* yielded mushrooms. The best casing medium for *C. india* during summer was FYM + sandy soil followed by soil + sand with bio efficiencies of 81 and 76%, respectively. In summer all the casing media gave satisfactory yield but during rainy season coir pith based media were not satisfactory. During rainy season soil + sand was the best medium with a bio efficiency of 52% (Table-4).

Table 2: Cultivation of *C. indica* in summer and rainy season in Kerala

Sl. No.	Substrates	Average spawn running (days)		Cropping cycle (days)		Days to first harvest		B.E.%	
		Summer	Rainy	Summer	Rainy	Summer	Rainy	Summer	Rainy
1	Paddy straw	20	24	57	62	30	35	75	52
2	Banana sheath	18	22	68	92	29	35	97	64
3	Rubber wood saw dust	44	48	91	98	68	76	10	8
4	Coir pith	29	30	89	94	48	55	26	18
5	Coconut saw dust	30	30	94	104	48	55	23	14
6	Portia tree saw dust	45	48	98	110	65	71	8	8

Cultivation of *T. georgii*

T. georgii was grown only on paddy straw and banana sheath and the other four substrates didn't give satisfactory spawn run and hence no yield. The best season for its cultivation was the S.W monsoon period in Kerala (June-August) (Table -3). In summer practicably there was no yield even on the best substrate. All the casing media except coir pith alone gave harvestable buttons in the rainy season. FYM + sandy soil and soil + sand gave the maximum yield during rainy season. In summer season none of the media could give harvestable mushrooms (Table-4). In summer season on paddy straw and banana sheath there was satisfactory spawn running and numerous pinheads developed after casing. But none of the pin heads could develop into harvestable buttons. The optimum conditions for development of *Tricholoma* buttons (plate 7 and 8) were found to be 22-25°C temperature and more than 90% R.H.

Table 3: Cultivation of *T. georgii* in rainy and summer season in Kerala

Sl. No.	Substrates	Spawn running period(days)		Cropping cycle (days)		Days to first harvest		B.E.%	
		Rainy	Summer	Rainy	Summer	Rainy	Summer	Rainy	Summer
1	Paddy straw	27	38	98	-	49	-	33	-
2	Banana sheath	23	35	114	-	46	-	42	-
3	Rubber wood saw dust	-	-	-	-	-	-	-	-
4	Coir pith	47	-	-	-	-	-	-	-
5	Coconut saw dust	46	-	-	-	-	-	-	-
6	Portia tree saw dust	-	-	-	-	-	-	-	-

Table 4: Effect of different casing media on the yield *of C. indica* and *T. georgii*

Sl. No.	Casing media	B.E. %			
		C.indica		T.georgii	
		Summer	Rainy	Summer	Rainy
1.	Soil + sand (3:1)	76	52	0	33
2	Coir pith + Soil (1:1)	64	30	0	28
3	FYM+sandy soil (1:1)	81	40	0	36
4	Coir pith + Vermi compost (1:1)	72	36	0	24
5	Coir pith alone	52	28	0	0

Cultivation of *G. lucidum*

Ganoderma pin heads were developed on all the four different type of substrates tried (plates 9, 10 and 11). Spawn running was satisfactory on all the media. The best medium for growing *G.lucidum* was found to be rubber wood saw dust enriched with rice bran (Table 5).

Table 5: Cultivation of *G. lucidum* on different substrates

Sl. No.	Substrates	Spawn running (days)	Days to first harvest	B.E.%	
				Rainy	Summer
1	Rubber wood saw dust	20	42	20	10
2	Portia tree saw dust	25	46	9	3
3	Coir pith	28	52	12	8
4	Coconut saw dust	30	56	8	5

All the five species of mushrooms were successfully cultured on all the media. But M.E.A and P.S.A. C.W. were the best media for the culture growth. For experimental purpose under the laboratory M.E.A would the best medium. But for ordinary growers who produce their own spawn, P.S.A. C.W. can be used for general culturing purposes since it is very cheap. The growth of mycelia of all the mushrooms especially *P. eous* and *P. sapidus* was very weak and feeble on P.D.A. and P.S.A. But their growth was very fast and thick on M.E.A. and P.S.A. C.W. This is the first report on the use of P.S.A. and P.S.A. C.W. for tissue culturing of any mushroom species. *P. eous* is the best variety to be cultivated during the period from January – March in Ghana with a biological efficiency of 57.4%.[2] *P. eous* is the perfect choice for the first time mushroom grower. It grows at room temperature and anywhere warmer. Ingale and Rmateke[3] reported that *P. eous* has the highest protein content of 46%, crude fiber 12% and highest mineral content among various *pleurotus spp.* The highest mineral content of *P. eous* makes it suitable for food supplement in diet. Telang *et al*[4] cultivated *P. eous* on different agro-waste substrates and maximum bio-efficiency of 82% was obtained on soybean straw. The mushroom was reported to contain 30.5% protein, 9% crude fiber and 6.5% ash. Sathyaprabha *et al*[5] reported the cultivation of *P. eous* on different agricultural waste substrates including paddy straw. In the present study also it was cultivated with a bio-efficiency of 50% on paddy straw in summer (January-May) in Kerala (Table-1). It was also found that

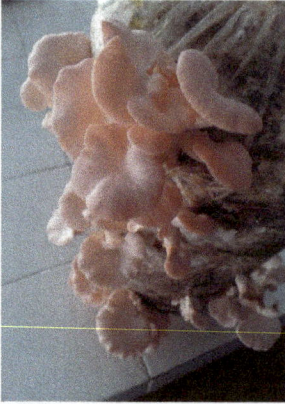

Plate-1

Plate-2

Plate-3

Plate-4

Plate-1   *P.eous* on paddy straw in the poly bag method of cultivation.

Plate-2   *P.eous* on Rubber wood saw dust in poly bag method of cultivation

Plate-3   *P.eous* on Rubber wood saw dust in plastic bottles

Plate-4   *P.sapidus* on paddy straw in plastic jar

Plate-5

Plate-6

Plate-7

Plate-8

Plate-5   *P.sapidus* on paddy straw in poly bag cultivation

Plate-6   *C.indica*, fully developed mushrooms on banana sheath

Plate-7   *T.georgii*, young mushroom buttons on banana sheath

Plate-8   *T.georgii*, fully developed mushrooms

Plate-9

Plate-10

Plate-11

Plate-9   *G.lucidum,* young mushrooms

Plate-10   *G.lucidum,* mushrooms ready for harvest

Plate-11   *G.lucidum,* mushrooms cultivated in a low cost mushroom shed

P. eous is the most ideal species for growing in plastic bottles and jars (plate 3,4). Protein analysis of the present strain revealed that it contains 52.5% protein whereas, the other *pleurotus spp.* contain about 25-30% protein. This particular strain was also found to be comparatively tolerant to different pests and diseases. This is the first report of cultivation of *p.eous* on rubber wood sawdust.

Similarly, *P. sapidus* was also an ideal species for growing on paddy straw and rubber wood sawdust throughout the year in the plains. It was grown on paddy straw with a bio efficiency of 40% (plate-5). On rubber wood saw dust the bio-efficiency was 60%. Telang *et al.* [6] reported successful cultivation of *P. sapidus* on different agro-waste substrates and maximum bio efficiency of 72.86% was obtained on soybean straw. The mushroom was reported to contain 26.75% protein, 7% ash, 52% carbohydrates, 2.6% fat and 91.33% moisture.

Mashandete[7] successfully cultivated *P. sapidus* on cat tail weed (*Typha domingesis*) substrate in Tanzania with a bio-efficiency of 100%. Singh and Singh [8] observed that the yield and bio efficiency of *P. sapidus* cultivated on substrates containing 20% and 30% vegetable wastes mixed with 70% and 80% paddy straw was better than that obtained on paddy straw alone. Bano and Rajarathnam[9] reported that *Pleurotus spp* have a wide range of temperature adaptability and substrates utilization. Mathew *et al* [10] evaluated three different substrates,viz., paddy straw, *Eliocharis plantogena* and rubber wood saw dust for the cultivation of five different species of *Pleurotus* in Kerala. *P. florida* was found to be the best species both in the plains as well as in the high ranges with 85% and 110% bio efficiency, respectively on paddy straw. The nutritional and medicinal values of the present isolate of *P. sapidus* have to be studied further.

The *C. indica* isolate under the present study could be successfully grown on paddy straw and banana sheath through out the year. Maximum bioefficiency of 92% was obtained on banana sheath and 75% on paddy straw during summer months. The corresponding bio efficiencies during rainy season were 64% and 52%, respectively. Artificial cultivation of milky mushroom was first reported by Purakayastha and Nayak and Purakayastha and Chandra.[11,12] Milky mushroom could be grown successfully on a wide range of agro-waste substrates like cereal straws, cotton etc.[13,14] Krishnamoorthy and Muthuswamy[15] evaluated different substrates for its cultivation and maximum yields were obtained on paddy straw and Maize stalks. Krishnamoorthy[16] observed that as it grows in hot humid climate the mushroom is highly suitable for cultivation in the Indian plains through out the year.

When compared to the existing isolate of milky mushroom cultivated in Kerala, the Kumarakom isolate is tastier and contains 42.5% protein on dry weight basis. Ram[17] reported that dried milky mushroom contained 32.3% protein and 64.2% starch. The new isolate contains 10% more protein. Nutritional and medicinal properties of this isolate need to be studied further.

T. georgii could be successfully grown on paddy straw and banana sheath with bio efficiencies of 33% and 42%, respectively (Table-3). This is an ecto- mycorrhizal mushroom and the best season for cultivation in the plains was the rainy season

(June-August). It was not suitable for cultivation in the high ranges. This is the first report of successful culturing, spawn production and cultivation of this mushroom and it contains 37.5% protein. The mushroom is creamy and light yellow in colour but very tasty and medicinal. Usually mycorrhizal mushrooms are difficult to be cultivated artificially. Bhavani Devi and Nair collected 5 ecto-mycorrhizal macro fungi in Kerala and one among them was identified as *T.georgii*.

Ebina *e al.*[18]studied the anti-tumor effect of peptide –glucan preparation extracted from the mycelium of pine mushroom (*Tricholoma matsutake*). Matsunaga *et al.*[19] standardized mass production of *T.matsutake* mycelia for its application to functional foods. Lin *et al.*[20] studied the free radical scavenging activity and inhibition of nitric acid production by four grades of *T. matsutake*. All these studies showed that *Tricholoma* are highly medicinal in nature. The exact medicinal properties of *T. georgii* need to be elucidated further.

For cultivating *C. indica* FYM+sandy soil and soil + sand were the best casing media with biological efficiencies of 81% and 76%, respectively. In the case of *T. georgii*, the corresponding biological efficiencies were 36% and 33%, respectively. Yadav[21] studied in detail the use of different vermi products in the cultivation of milky mushroom. In the present study coir pith + vermicompost (1:1) gave 72% and 24% bio efficiency, respectively in the case of *C. india* and *T. georgii* when it was used as the casing medium.

G. lucidum was cultivated successfully on rubber wood saw dust- rice bran medium for the first time in Kerala (plate 9, 10). A maximum bio efficiency of 20% was obtained during rainy season followed by 10% during summer. It is known as the King of medicinal mushrooms and recommended as a neutraceutical food supplement for the management of several health problems. Tiwari *et al.*[22] reported successful cultivation of G. *lucidum* on saw dusts of broad leaved trees enriched with wheat bran. They also reported that *G. lucidum* contains high amounts of Calcium and high Potassium/Sodium ratio. The Potassium/Sodium ratio in G. *lucidum* is higher than all the other species of mushrooms and hence, beneficial for vascular problems. Veena and Pandey[23] reported the cultivation of *G.lucidum* in Kerala on paddy straw + sawdust + rice bran (22.5 + 67.5 + 10) with a bio efficiency of 29.9%. Rai *et al.* [24] reported successful cultivation of *G.lucidum* on saw dusts of different broad leaved trees (Mango, Poplar, Coconut and Seesham). Sharma and Sathishkumar [25] reported that extracts from fruit bodies and mycelia of *G. lucidum* occurring in South India were found to possess *in vitro* anti-oxidant property. It is the most popular medicinal mushroom in China and has been used for a wide range of health benefits over the past 4000 years. It is known for managing cancer in combination with conventional therapy and also for its anti HIV effect.

G. lucidum is the most pharmacologically and commercially important medicinal mushroom in the world with current annual global trade of about two billion dollars. Trade in India has crossed Rs.100 crores annually through import from Malaysia and China.[25] The medicinal values of Indian Ganodermas need to be studied further. However, the wound healing property was tested in the present study and found that it was very effective in healing mechanical wounds in human body. Since, its cultivation has been standardized we can think of producing it in our country and make it beneficial to a large section of the population at a cheaper rate. Some workers consider *G. lucidum* as a mild pathogen to several forest trees and crops like coconut and areca nut. In Kerala it is responsible for causing foot rot and wilt in coconut and areca nut. Hence, it is important that due care is taken during cultivation and in the disposal of spent substrate. It should be cultivated under fully controlled conditions so that none of the spores are escaped to outside. The spent substrate could be dried and burnt as fuel.

References

1. Garcha, H.S, Khann, P.K. and Soni, G.L., Nutritional importance of mushrooms. In mushroom biology and mushroom products. The Chinese University Press (Chang, S.T; Buswell, J.A and Chiu, S. (Eds). 1993, 227-235.

2. Obodai, M. and Vowotor, K.A., Performance of different strains of *Pleurotus* species under Ghanaian conditions. *J Food Technol Africa*. 2002, **7**, 98-100.

3. Ingale, A. and Ramteke, A., Studies on cultivation and Biological efficacy of mushrooms grown on different agro residues. *Innovative Romanian Food Biotechnol*, 2010,**6**,25-28

4. Telang,S.M;Patil,S.S. and Baig, M.M.V., Biological efficiency and nutritional value of *Pleurotus sapidus* cultivated on different substrates. *Food Sci Res J*. 2010,1,127-29.

5. Sathyaprabha, G,, Paneerselvam, A. and Kumaravel, S., Cultivation of *Pleurotus platypus* and *Pleurotus eous* in different agricultural waste substrates. *J Pharm Res*. 2011,**4**, 2543-44.

6. Telang, S.M., Patil, S.S. and Baig,M.M.V., Comparative study on yield and nutritional aspect of *Pleurotus eous* mushroom cultivated on different substrates. *Food Sci Res J*., 2010,**1**,60-63.

7. Mashandete, A.M., Cultivation of *pleurotus* H.K-37 and *pleurotus sapidus* on cat tail weed (*Typha domingesis*) substrate in Tanzania. *Int J Res Biol Sci*, 2011, **I**, 35-44.

8. Singh, M.P. and Singh, V.K., Biodegradation of vegetable and agro-wastes by *Pleurotus sapidus*: A novel strategy to produce mushroom with enhanced yield and nutrition. *Cell Mole Biol*, 2012,**58**,1-7.

9. Bano, Z. and Rajarathnam, K., *Pleurotus* mushroom as a nutritious food. In: Tropical mushrooms: Biological nature and cultivation methods. The Chinese University Press, Chang S.T and Quimio, T.H. (Eds.)1982.

10. Mathew, A.V., Mathai, G. and Suharban, M., Performance evaluation of five different species of *Pleurotus* (Oyster mushroom) in Kerala. *Mushr Res*. 1986,**5**,9-12.

11. Purakayastha, R.P. and Nayak, D.K., Studies on the production and qualitative improvement of *Calocybe indica*. *Mushr Sci*, 1978,**1**, 439-44.

12. Purakayastha, R.P. and Chandra, A., Manual of Indian edible mushrooms. Today & Tomorrow printers and publishers, New Delhi. 1985,192-94.

13. Trivedi, A, Sharma, S.S. and Doshi, A., Cultivation of *Calocybe indica* under semi-arid conditions. In: Indian Mushrooms, Nair, M.C., (Ed.) Kerala Agriculture University, Vellanikkara, Kerala, 1991, pp166-168.

14. Rajarathnam, S., Shashirekha, M.V. and Bano, Z., Biopotentialities of the Basidio-macromyectes. *Adv Appl Microbiol*, 1993, **37**, 233-61.

15. Krishnamoorthy, A.S. and Muthuswamy, M., Milky mushroom. *Kisan World*, 1997, 39

16. Krishnamoorthy, A.S., Commercial prospects of milky mushroom (*Calocybe indica*) in the tropical plains of India. In: current vistas in mushroom biology and production Upadhyay. R.C; Singh, S.K. and Rai. R.D. (Eds.), 2004.

17. Ram, R.C., Cultivation of milky mushroom. *Kisan World*, 2004,55.

18. Ebina, T., Kubota, T., Ogamo, N. and Matsunaga, K., Anti-tumor effect of a peptide – glucan preparation extracted from mycelium of *Tricholoma matsutake*. *Biother*, 2002,**16**, 255-59.

19. Matsunaga, K., Chiba, T. and Takahashi, E., Mass production of Matsutake (*Tricholoma matsutake*) mycelia and its application to functional foods. *Bioidustry*, 2003,**20**, 37-46.

20. Lin, H.W., Yoon, J.H., Kim, Y.S., Lee, M.W., Park, S.Y. and Choi, H.K., Free radical scavenging and inhibition of nitric oxide production by four grades of pine mushroom (*Tricholoma matsutake*). Food Chem, 2007, **103**,1337 – 1342.

21. Yadav, R.S., Use of vermiproducts in the cultivation of Milky mushroom (*Calocybe indica*). M.Sc (Ag) thesis submitted to the University of Agricultural Sciences, Dharward.2006.

22. Tiwari,C.K., Meshram, P.B. and Patra, A.K., Artificial Cultivation of *Ganoderma lucidum. The Indian Forester*, 2004,**130**, 9.

23. Veena, S.S. and Pandey, M., Paddy straw as a substrate for the cultivation of Lingzhi or Reishi Medicinal mushroom. *Ganoderma lucidum* (W.Curt.Fr) P. Karst. I-in India. *Int. J. Med. Mushr.*, 2011,13, 397-400.

24. Rai, R.D., Singh, S.K., Yadav, M.C. and Tewari, R.P., Mushroom Biology and Biotechnology. N R C M., Solan. 2007, pp 388.

25. Sharma, V.P. and Sathishkumar., Cultivation of medicinal mushroom – *Ganoderma lucidum* . In: Mushrooms cultivation, marketing and consumption. Singh, M., Vijay, B., Kamal, S. and Wakchaure, G.C. (Eds.), 2011, pp. 254.

14

Ganoderma lucidum (Fr.) P. Karst: A promising target to ameliorate mitochondrial dysfunction in Alzheimer's disease?

T.A. Ajith*[1], N.P. Sudheesh[2] and K.K. Janardhanan[3]

[1]Department of Biochemistry, Amala Institute of Medical Sciences, Amala Nagar, Thrissur-680 555, Kerala, India

[2]Technology Development and Transfer (TDT) Division, Dept. of Science & Technology, Ministry of Science & Technology, New Delhi 110 016

[3]Department of Microbiology, Amala Cancer Research Centre, Amala Nagar, Thrissur-680 555, Kerala, India.*email-taajith@rediffmail.com

Pathogenesis of neurodegenerative diseases such as Alzheimer's disease, and Parkinson's disease, may involve the generation of reactive oxygen species (ROS) and mitochondrial dysfunction. Defects in the activities of complexes of mitochondrial electron transport chain (ETC), possibly associated with oxidant/antioxidant imbalance, are thought to underlie the defects in energy metabolism that induce cellular degeneration. In Alzheimer's disease (AD), the defect in energy metabolism was found to be severe. Hence, supplementation with antioxidants such as lipoic acid in patients with AD resulted in mild cognitive improvements probably mediated by protecting mitochondria against respiration-linked oxidative stress. The results of the previous studies concluded that the extract of *Ganoderma lucidum* was effective to improve either the cellular energy status or the age-related decline of antioxidant status in post mitotic tissues. Beneficial effects of extract of *G. lucidum* on the activities of mitochondrial dehydrogenases; complex I and II of ETC have been demonstrated in the aged rat brain. Further, its antioxidant and anti-inflammatory activities have been established. Hence, accounting the various biological activities, *G. lucidum* can be used as a supplementary agent to alleviate the inflammation, oxidative stress as well as the mitochondrial impairment in the brain that may delay the progress of ageing associated human ailments. This article discusses the role of mitochondria in the pathophysiology of AD and the possible role of *G. lucidum* as an adjuvant to other therapeutic intervention.

Introduction

Mitochondrial dysfunction has been associated with a wide variety of human diseases. Today, this energy producing cellular organelle is involved in more than thousands of human diseases. Intensive research during the last few decades elucidated the role of mitochondria in neurodegenerative diseases such as Parkinson's disease, Alzheimer's disease, Friedreich's ataxia, multiple sclerosis and amyotrophic lateral sclerosis.[1] The role of mitochondiria in these diseases is supported by the mitochondrial cascade hypothesis which states that somatic mutations in mitochondrial DNA (mtDNA) causes energy failure and reinforces mtDNA damage through the excess formation of reactive oxygen species (ROS).[2] Though the physiological level of ROS especially hydrogen peroxide has been involved in the signal transduction pathways that result in the cell differentiation, immune cell activation and metabolic adaptation of cells in hypoxic condition, their excess production beyond the physiological limit can evoke oxidative stress[3]. The major factor in the dysfunction of mitochondria has been hypothesized about the defects in oxidative phosphorylation (OXPHOS) and the stimulation of mitochondrial production of ROS and damage to mt DNA.[1,4] Vicious cycle that operates from the oxidative stress condition in the mitochondria is depicted in figure 1.

Alzheimer's disease AD is a fatal neurodegenerative disorder identified in 1901 by German Psychiatrist Aloi Alzheimer in a fifty-year-old woman. In spite of more than 100 years of research, the exact mechanism underlying this incurable disease is not yet been established. Deposition of amyloid-β (Aβ) is one of the causative agents of AD pathology followed by neurofibrillary tangles (NFT), (aggregates of tau), cell loss, vascular damage and dementia[5]. The formation of amyloid plaques (aggregates of Aβ) and NFT is accompanied by mitochondrial dysfunction, but the mechanisms underlying this dysfunction are poorly understood. Mitochondrial malfunction has been reported as the primary pathology in AD.[5,6]

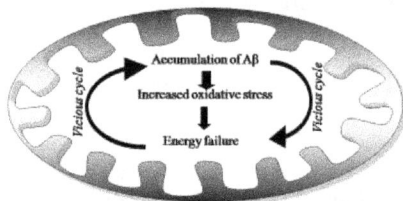

Fig. 1: Mitochondrial Aβ generation results in vicious cycle of mitochondrial functional decline

Medicinal mushrooms represent a major untapped source of powerful group of pharmaceutical agents that remain to be explored. Evaluations of therapeutic application and clinical use of medicinal mushrooms occurring in India has not received adequate attention. In China, the medicinal mushroom has been considered as a health supporting traditional medicine. Investigations carried out at Amala Cancer Research Centre during the last decade can be considered as the major contribution in this area. *Ganoderma lucidum* (Fr) Karst is a wood inhabiting macrofungus, commonly known as Reishi or Ling Zhi. In Chinese folklore, fruiting bodies of *Ganoderma* have been regarded as panacea for all type of diseases. It carries a diverse set of bioactive compounds. We have investigated the pharmacological properties of extracts from *G. lucidum* and the results showed that extracts of the fruiting bodies of *G. lucidum* occurring in South India possessed profound antioxidant, anti-inflammatory and antitumor properties.[7-9] The anticancer and radioprotective effects of this mushroom has also been reported.[10,11] Since, free radicals are involved in the pathogenesis of most of the human ailments compounds with antioxidant activity may used as a prophylactic agent against such diseases. This article discusses the possible therapeutic role of *G. lucidum* in AD.

Role of mitochondria in the formation of amyloid-β

Amyloid β precursor protein (APP), protein produced by gene located on the chromosome 21, is critical to neuron growth, survival and post-injury repair.[12,13] It is mainly involved in the synaptic formation and repair. APP is commonly inactivated by proteases (belong to the family of secreatase ie. α and β secreatase) mediated cleavage of extra cellular fraction and further apoptosis of the released membrane bound fraction [14]. The mitochondrial APP levels increase with the severity of AD disease. The targeting of APP to mitochondria is taken as an AD specific process. The role of mitochondria in the generation of Aβ is ascribed to the targeting of APP. A dual leader sequence in the newly synthesized APP has permitted its targeting either to the endoplasmic reticulum (ER) or to mitochondria[15]. Mitochondrial APP forms complexes with the protein importation translocases of the outer (TOM) as well as that of the inner membranes (TIM), resulted in the reduced importation of respiratory chain subunits and other mitochondrial proteins.[15] This eventually decreases the activity of respiratory chain enzymes and increase of free radical generation. Furthermore, the clogged APP in the mitochondrion is subjected to amyloidogenic processing by gamma secretase and forms the protein Aβ.[15]

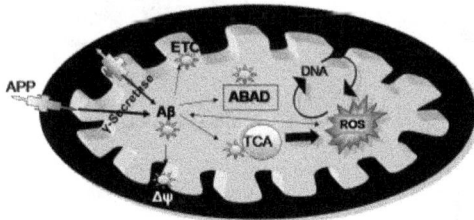

Fig. 2: Formation of Aβ and its effect on mitochondrion

Aβ protein can trigger mitochondrial dysfunction through a number of pathways, such as impairment of OXPHOS, elevation of ROS production, alteration of mitochondrial dynamics, and interaction with mitochondrial proteins.[16] Mitochodrial impairment due to the formation of Aβ protein is given in figure 2. Aβ formed in the mitochondria interacts with amyloid-binding alcohol dehydrogenase (ABAD, peptide which can metabolize a broad array of substrates, including alcohols and ketone bodies in ATP depleted state) and produces ROS.[17] Aβ also inhibits activities of cytochrome oxidase of ETC and the α-ketoglutarate dehydrogenase of the TCA cycle.[17] Declined amyloid degrading enzymes, such as insulin degrading enzyme (IDE) and presequence peptidase (PreP), are also found in mitochondria.[18] This may result in the hyperphosphorylation of proteins involved in the antiinsulin hormones mediated pathways in the cells of brain. The role of one of such proteins, Tau in the formation of neurofibrillary tangels and their aggregation results in the formation of neuritic plaque has been reported in the autopsy samples of AD patient. .

Mitochondrial impairment

Synaptic dysfunction and the loss of synapses are the early pathological features of AD. Aβ protein induced impairment of OXPHOS trigger the release of ROS which is involved in several vicious cycles that results in the damage of mitochondrial DNA and also in the generation of more and more Aβ protein.[19] The neurons in the AD brain are chronically exposed to excessive amounts of H_2O_2 and hydroxyl radicals due to mitochondrial defects. Declined ATP production will further elevate the cytosolic calcium levels owing to leakage from Aβ protein modified calcium-conducting channels in the ER and plasma membrane.[20] Cross talk between calcium over load and ROS generation has been described (Figure 3). The elevated cytosolic calcium may result in calcium overload of mitochondria and deregulate the neuronal calcium signaling. In such an energy deficit state, release of proapoptotic proteins from damaged mitochondria results in apoptosis (Figure 4). Synaptic mitochondria showed a greater degree of age-dependent accumulation of Aβ and mitochondrial alterations.[21] At low concentrations (low-nanomolar) Aβ remains monomeric and can function as an antioxidant. However, at higher concentrations, aggregation of Aβ produces H_2O_2.[22] Oligomeric Aβ is viewed as the most toxic.[23]

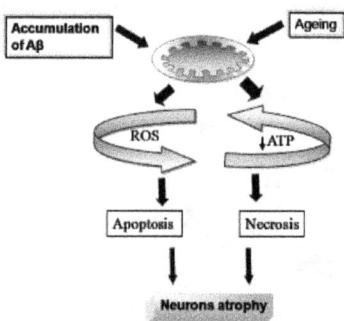

Fig. 3: Schematic of ROS-induced apoptosis and necrosis   **Fig. 4:** Role of Aβ in the influx of calcium into the cytosol and associated neuronal injury

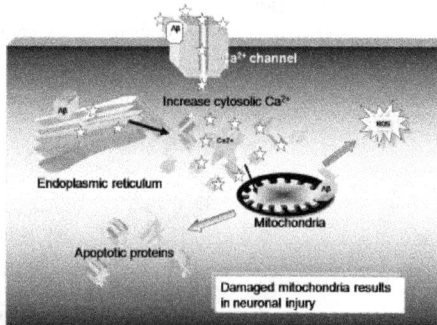

The released ROS, neurotoxic cytokines and proteases from the microglial cells, primary cells in the innate immune response in brain, can evoke an inflammatory response in the adjacent astrocytes, the glial cells that supply neutrients including

cholesterol and fats to neurons.[14] The oxidative stress irreversibly damages the proteasome and the lysosome systems in the microglial cells. This mitochondrial-mediated impairment of autophagy will potentiate the Aβ deposition. Mitochondrial malfunction starts sequences of events leading to NAD+ induced sirtuin-2, a class of proteins that regulate wide range of cellular processes including the mitochondrial biogenesis, activation and microtubule network breakdown.[5]

All these effects will either result in the atrophy or increase death of neuron and eventually unable to function the neurons in their intended role ie. neurotransmission. AD brains are chronically exposed to excessive amounts of glutamate, a neurotransmitter as well as an oxidative agent. The oligomeric intracellular Aβ protein can also produce NFT and neuritic plaque, further enhances the apoptotic cell death. A genetic defect in mitochondrial Complex I genes is associated with a small minority (2%) of AD cases.[24] However, the most documented reduction of mitochondrial enzyme activity in AD is the activity of complex IV.[25]

Treatment of AD

Researchers have been focused on the pharmacological therapies that interfere with the synthesis of Aβ, but many are faced disappointing results. Acetylcholinesterase inhibitors (tacrine, rivastigmine, galantamine and donepezil) and the others such as memantine, an NMDA receptor antagonist, are currently used approved medicines.[26, 27] These agents can also be used in combination. Conclusions of various prospective studies using these agents alone and in combination are debatable and inconclusive and also some of them are marginally effective. Therapies using these agents are not free from side effects, in approximately 10–20% of users are mild to moderate in severity of muscle cramps, bradycardia, decreased appetite and weight, and increased gastric acid.[28] Many clinical and experimental trials using compounds with the anti-inflammatory and antioxidants are reported. Use of an experimental vaccine was found to clear the amyloid plaques in early human trials, but it did not have any significant effect on dementia.[29] Early diagnosis of AD is difficult, therefore an important challenge for the successful management of AD is either the development of new tools to detect AD in its earliest stages which could predict the progression of the disease or select a therapeutic regimen to slow down the progression of the disease.

Effect of *G. lucidum* on the energy status of the mitochondria in the aged rat brain

Ethanol extract of *Ganoderma lucidum* on the activities of mitochondrial dehydrogenases; complex I and II of electron transport chain have been evaluated in the aged rat brain. Aged male Wistar rats were administered with ethanol extract of *G. lucidum* (50 and 250 mg/kg, p.o) once daily for 15 days. Results of the study demonstrated that the extract of *G. lucidum* significantly ($p < 0.01$) enhanced the activities of succinate dehydrogenase, malate dehydrogenase, α-ketoglutarate dehydrogenase, pyruvate dehydrogenase, complex I and II when compared to that of the aged control animals. The level of the lipid peroxidation was significantly lowered ($p < 0.01$) in the *G. lucidum* treated group. However, no statistically significant difference could find between the activities of enzymes in groups treated with 50 and 250 mg/kg of *G. lucidum*.[30] Further, the effect of *G. lucidum* on the antioxidant status in the mitochondria of brain of aged mice has been evaluated. Oral administration of ethanol extract of *G. lucidum* (50 and 250mg/kg), once daily for 15 days significantly (p < 0.05) elevated the levels of activities of manganese-superoxide dismutase, glutathione peroxidase, glutathione-*S*-transferase and catalase as well as levels of reduced glutathione. Lipid peroxidation, advanced oxidation protein products and ROS levels were found to be decreased.[31]

Possible role of *G. lucidum* in Alzheimer's disease

It is evident that cumulative oxidative damage over the lifespan of an organism can affect mitochondrial efficiency and leave neurons susceptible to cell death. Brain is more vulnerable to the ROS induced damages since it has very low antioxidant status and more polyunsaturated fatty acids.[32] Accounting the excess generation and associated oxidative stress in mitochondria, agents which can ameliorate the oxidative stress and protect the mitochondria from the damages can be suggested as prophylactic therapeutic agent. Chronic increases in oxygen radical production in the mitochondria can lead to a catastrophic cycle of mt DNA damage as well as functional decline, further oxygen radical generation, and cellular injury.[33] Agents with additional biological activities such as anti-inflammatory effect can further strengthen their possible role. Experimental evidences during the last decade at our centre reported various pharmacological properties of this macro fungus. Among them, its effect in aged rat brain to enhance the activities of mitochondrial dehydrogenases; complex I and II of electron transport chain need special attention.

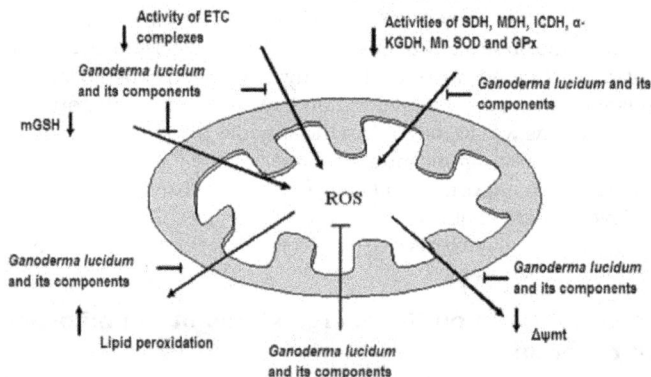

Fig. 5: Effect of *G. lucidum* in mitochondria

Reduction in cholesterol bioavailability would lead to reduced neurotransmission at the synapse. Deficiency in the supply of cholesterol, fats, and antioxidants to the brain has been reported in many research reports. Neurons begin to suffer damage as well, once the astrocytes can no longer supply adequate amounts of cholesterol, fats and antioxidants. Cardiovascular risk factors, such as hypercholesterolaemia, hypertension, diabetes, and smoking, are associated with a higher risk of onset and course of AD.[32] Ganoderma species are enriched sources of fatty acids and phytochemicals mainly terpenoids to protect the cell from the free radical damage. More than 150 terpenoids in the *G. lucidum* so far reported and many of them may reduce the generated ROS.[34] Correlating the various biological activities of *G. lucidum*, it may be postulated the possible therapeutic role in AD with other pharmacological interventions but need more detailed scientific evidences (Figure 5).

Conclusion

Mitochondrial oxidative stress has been demonstrated in the brain of AD patients. The altered mitochondrial dynamics may lead to decreased mtDNA copy number and hence the mitochondrial electron transport activity. Therefore, preserving the mitochondria in the brain of human in old age will be one of the appropriate therapeutic interventions to extend the incidence of AD. Antioxidants and anti-inflammatory agents that are used for the treatment of AD produce an inconclusive beneficial out come. However, the standard treatment using acetyl choline esterase inhibitors at the early stage produce an extension in the progression of the disease. Hence, it could be effective if compounds with multiple actives such as antioxidants and anti-inflammatory used along with the standard regimen. In light of our previous work, especially the efficiency of *G. lucidum* in increasing the energy status in the brain mitochondria of aged rat, it can be suggested the prophylactic use of this mushroom in AD supplemented with the standard medical interventions.

References

1. Santo, R.X., Correia, S.C., Wang, X.X., Perry, G., Smith, M.A., Moreira,P.1. Zhu, X., Alzheimer's Disease: Diverse Aspects of Mitochondrial Malfunction, *Int. J. Clin. Exp. Pathol.*, 2010, **3**, 570–581.

2. *Hardy, J.A, Higgins G.A., Alzheimer's disease: the amyloid cascade hypothesis. Science. 1992,***256***,184–185.*

3. Smith, M.A., Rottkamp, C.A., Nunomura, A., Raina, A., Perry, G., Oxidative stress in Alzheimer's disease. *Biochim. Biophys. Acta*, 2000,**1502**,139–144.

4. Hutchin, T.P., Heath, P.R., Pearson, R.C., Sinclair, A.J., Mitochondrial DNA mutations in Alzheimer's disease. *Biochem Biophys Res Commun.*, 1997,**241**,221–225.

5. Diana, F.F., Silva Esteves, A.R., Oliveira, C.R., Cardoso, S.M., Mitochondria: The Common Upstream Driver of Abeta and Tau Pathology in Alzheimer's Disease. *Proc. Natl. Acad. Sci. USA*, 2010,**107**,18670-18675.

6. Beal, M.F., Mitochondrial dysfunction in neurodegenerative diseases. *Biochimica. Et. Biophysica. Acta.*,1998,**1366**,211–223.

7. Lakshmi, B., Ajith, T.A., Sheena, N., Gunapalan, N., Janardhanan, K.K., Antiperoxidative, anti-inflammatory, and antimutagenic activities of ethanol extract of the mycelium of *Ganoderma lucidum* occurring in South India. *Teratog. Carcinog. Mutagen.*, 2003;Suppl 1,85-97.

8. Jones, S., Janardhanan, K.K., Antioxidant and Antitumour activity of Ganoderma lucidum (Curt.:Fr) P.Karst. –Reishi (Aphyllophoromycetideae) from South India, *Int. J. Med. Mushr.*, 2000, **2**, 195-200.

9. Lakshmi, B., Ajith, T.A., Jose, N., Janardhanan, K.K., Antimutagenic activity of methanolic extract of Ganoderma lucidum and its effect on hepatic damage caused by benzo[a]pyrene. *J Ethnopharmacol.* 2006,**107**,297-303.

10. Lakshmi, B., Sheena, N., Janardhanan, K. K., Prevention of mammary adenocarcinoma and skin tumour by *Ganoderma lucidum*, a medicinal mushroom occurring in South India. *Cur. Sci.,* 2009, **97**,1658-1664.

11. Pillai T.G., Nair, C.K.K., Janardhanan, K.K., Enhancement of repair of radiation induced DNA strand breaks in human cells by *Ganoderma* mushroom polysaccharides. *Food Chem.,* 2010,**119**,1040–1043.

12. Priller, C., Bauer,T., Mitteregger, G., Krebs, B., Kretzschmar, H.A., Herms, J., Synapse Formation and Function is Modulated by the Amyloid Precursor Protein. *J. Neurosci..* 2006,**26**,7212–221.

13. Turner,P.R., O'Connor, K., Tate, W.P., Abraham, W.C., Roles of Amyloid Precursor Protein and its Fragments in Regulating Neural Activity, Plasticity and Memory. *Prog. Neurobiol..* 2003,**70**,1–32.

14. Querfurth, H.W., LaFerla, F.M., Alzheimer's disease. *N. Engl. J. Med.,* 2010;**362**:329-344

15. Lin, M.T., Beal, M.F., Alzheimer's APP mangles mitochondria. *Nature Med.,2006,* **12**, 1241 – 1243

16. Moreira, P.I., Carvalho, C., Zhu, X., Smith, M.A., Perry, G., Mitochondrial dysfunction is a trigger of Alzheimer's disease pathophysiology. *Biochim. Biophys. Acta.,*2010,**1802**,2-10

17. Hirai, K., Aliev, G., Nunomura, A., Fujioka, H., Russell, R.L., Atwood, C.S., Johnson, A.B., Kress, Y., Vinters, H.V., Tabaton, M., Shimohama, S., Cash, A.D., Siedlak, S.L., Harris, P.L.R., Jones, P.K., Petersen, R.B., Perry, G., Smith, M.A., Mitochondrial abnormalities in Alzheimer's disease. *J. Neurosci.* 2001,**21**,3017–3023.

18. Reddy, P.H., Manczak, M., Mao, P., Calkins, M.J., Reddy A.P., Shirende,U., Amyloid-β and Mitochondria in Aging and Alzheimer's Disease: Implications for Synaptic Damage and Cognitive Decline, *J Alzheimer's Disease,*2010, **20**, S499–S512.

19. Mancuso, M., Orsucci, D., Siciliano, G., Murri, L., Mitochondria, mitochondrial DNA and Alzheimer's disease. What comes first?. *Curr. Alzheimer Res.,* 2008, **5**,457-568

20. Starkov, A.A., Beal,F.M., Portal to Alzheimer's disease. *Nat .Med.* 2008, **14**,1020–1021.

21. Du, H., Guo, L., Yan, S., Sosunov, A.A., McKhann, G.M., Yan, S.S., Early deficits in synaptic mitochondria in an Alzheimer's disease mouse model. *Proc. Natl Acad.Sci., USA,*2010,**11**,1193-1206.

22. Rhein, V., Baysang, G., Rao, S., et al., Amyloid-beta leads to impaired cellular respiration, energy production and mitochondrial electron chain complex activities in human neuroblastoma cells, *Cell. Mole. Neurobiol.,*2009, **29**, 1063–1071.

23. Lesné, S., Koh, M.T., Kotilinek,L., Kayed, R., Glabe, C.G., Yang, A., Gallagher, M., Ashe, K.H. A specific amyloid-beta protein assembly in the brain impairs memory. *Nature*, 2006, **440**,352-357.

24. Chapira, A.H., Cooper, J.M., Dexter, D., Clark, J.B., Jenner, P., Marsden, C.D., Mitochondrial complex I deficiency in Parkinson's disease. *J. Neurochem.*, 1990,**4**,823-827.

25. Maurer, I., Zierz, S., M"oller, H.J., A selective defect of cytochrome c oxidase is present in brain of Alzheimer disease patients, *Neurobiol. Aging*, 2000,**21**,455–462.

26. Pohanka, M., Cholinesterases, a target of pharmacology and toxicology. *Biomedical Papers Olomouc* 2011, **155**, 219–229.

27. Muir, K.W., Glutamate-based therapeutic approaches: clinical trials with NMDA antagonists. *Curr Opinion in Pharmacol* 2005,**6**, 53–60.

28. Birks, J., Harvey, R,J,, Donepezil for dementia due to Alzheimer's disease. In Birks, Jacqueline. *Cochrane Database Syst Rev* 2006-01-25 (1).

29. Holmes, C., Long-term Effects of Abeta42 Immunisation in Alzheimer's Disease: Follow-up of a Randomised, Placebo-controlled Phase I Trial. *Lancet.* 2008,**372**,216–223.

30. Ajith, T.A., Sudheesh N.P., Roshny, D., Abishek G., Janardhanan, K.K., Effect of *Ganoderma lucidum* on the activities of mitochondrial dehydrogenases and complex I and II of electron transport chain in the brain of aged rats. *Exp. Gerontol.* 2009,**44**, 219–223.

31. Sudheesh, N.P., Ajith, T.A., Ramnath, V., Janardhanan, K.K., Therapeutic potential of *Ganoderma lucidum* (Fr.) P. Karst. against the declined antioxidant status in the mitochondria of post-mitotic tissues of aged mice. *Clin Nutr.* 2010,**29**,406-412.

32. Blomgren, K., Hagberg H., Free radicals, mitochondria, and hypoxia–ischemia in the developing brain. *Free Radic. Biol. Med.*2006, **40**, 388 – 397.

33. Ajith, T.A., Mitochondria: oxidative stress, dysfunction and cell death. *VNT Biol. Med. Chem.*, 2013,**1**,21-32.

34. Chen, S., Xu, J., Liu C et al., Genome sequence of the model medicinal mushroom *Ganoderma lucidum. Nature Commun.*, 2012,**3**,1-9

15

Diversity and conservation of medicinal mushrooms of India

Meera Pandey[1*] and **S.S Veena**[2]

[1]Mushroom Research Laboratory, Indian Institute of Horticultural Research Hessaraghatta lake post, Bangalore-560089, India.

[2]Division of crop protection, CTRI, Sreekariyam, Thiruvananthpuram -695017, Kerala, India. *E-mail: meera@iihr.ernet.in; meerapandeyiihr@gmail.com

Wild mushroom gathering is an important economical activity in many parts of India and has become a sustainable source of income, nutrition and as medicinal herbs. Thanks to the pharmaceutical and clinical research undertaken over the last 50 years; mushrooms are now revered as important functional foods. One of the most important requirements for sustaining the medicinal mushroom industry is the conservation and research on the mushroom diversity of the country. The present work encompasses the documentation of medicinal mushrooms of the tropical and subtropical forests of Western ghats (Karnataka), Aravalli forest range of Rajasthan, Gir and Ahwa forest region of Gujarat, Tropical forests of Sikkim, Tropical forests of South and Middle Andaman and the Balaghat and Jabalpur forest range of Madhya Pradesh. Many of the species were cultured and conserved in the germplasm repository of Indian Institute of Horticultural Research, Bangalore and are being investigated for their domestication and commercialization.

Introduction

Mushrooms have long been associated with human civilization. Wild mushroom gathering is an important economical activity in many parts of India and has become a sustainable source of income and nutrition. Ethno mycological records of China and Australia reveal the usage of mushrooms for healing and other medicinal effects. As many as 700 mushroom species are known to have medicinal properties. Global medicinal mushroom trade was an estimated US $ 14 billion in 2000. A whole new array of mushroom nutriceutical industries has been established. Thanks to the pharmaceutical and clinical research undertaken over the last 50 years; mushrooms are now revered as important functional foods. A myriad of texts and reports have been published showing concern over conservation of genetic resources, destruction of forests, extinction of species and the effects of global warming. Most of such works however lacks content on fungi or any micro-organisms. Fungi in general and mushrooms in particular have received scant attention in the biodiversity debate due to a lack of awareness amongst biologists of their significance role in nutrition, medicine, in ecosystem function and in human progress. The time has now come to broaden the biodiversity debate by focusing on its fungal dimension[1]. The alarming rate of destruction of tropical and temperate mushroom habitats not yet explored makes mushroom conservation a key issue for Indian Mushroom research. A whole new unexplored wealth may be discovered in our forests. The present work encompasses the documentation of medicinal mushrooms and local knowledge of the tropical and subtropical forests of Western ghats (Karnataka), Aravalli forest range of Rajasthan, Gir and Ahwa forest region of Gujarat, Tropical forests of Sikkim, Tropical forests of South and Middle Andaman and the Balaghat and Jabalpur forest range of Madhya Pradesh. Many of these species have been cultured and conserved in the germplasm repository of Indian Institute of Horticultural Research, Bangalore and are being investigated for their domestication and commercialization.

Experimental

Explorations were undertaken during the months of August and September from 2005 to 2012. The regions of explorations included Western ghats (Shimoga, Kodagu and Dakshin Kanada districts), Rajasthan (Forests in Udaipur, Sirohi, Rajsamand, Chitoorgarh, Banswara, Jhalawar, Sawai Madhopur and Jodhpur), Gujarat (Ahwa, Junagad and Bhuj), Sikkim (East, West, South and north Sikkim), Madhya Pradesh (Balaghat, Jabalpur) and A&N islands (South Andaman, Rutland island, Neil island, Havelock island and Middle Andaman). The forests explored ranged from tropical wet evergreen forest, tropical semi evergreen forest, moist deciduous forest, dry deciduous forest and tropical thorn and scrub forests.

The passport data of the documented species was generated. The species were first identified on the basis of sporophores characters[2] and some of them cultured at the spot of collection. The spore print of the species was also made to aid in identification and culturing. The cultures were then purified in the laboratory and validated through induction of fructification. Tissue cultures were raised from laboratory induced sporophores and these validated cultures were characterized in terms of mycelial characters, spore characters and optimal growth requirements for further conservation.

Spawn was prepared by the conventional technique on sorghum grains. The induction of sporophore was done on pasteurized paddy straw (80-85°C for 2 hours) or on sterilized sawdust (15 lb pressure, 121°C, 1 hour). The fructification validated the authenticity of the original culture. Tissue cultures were raised from laboratory grown sporophores and conserved. Sporophore morphology, basidiospore morphology and other specific traits like production of dye etc was also studied. The cultures are being maintained in the mushroom culture repository of IIHR.

A total of 268 species were documented of which 72 species are known to be medicinal. Some of the important culinary medicinal/ medicinal mushrooms documented/ conserved are as follows.

Pleurotus species: Commonly called as oyster mushrooms, these species constitutes 25% of the global mushroom market. Four indigenous species have been collected by IIHR (forests of Western ghats of Karnataka, forests of Balaghat in Madhya Pradesh and forests of Andaman & Nicobar islands) and have been conserved at IIHR mushroom germplasm repository. *P. djamor*, collected from Western ghats has been domesticated and commercialized by IIHR as Arka-OM-1 and its spawn is available at IIHR for commercial cultivation. *P. cystidiosus* (abalone oyster mushroom) is in the pipeline for commercialization. A white and another pink *Pleurotus spp.* (species yet to be identified) collected from Balaghat forests (MP) are being studied for domestication and commercialization.

Regions of documentation from India: Western ghats (Shimoga), A&N islands (port blair), Madhya Pradesh (Balaghat), Bangalore

Medicinal properties: *Pleurotus* species have shown anti-tumor, cholesterol reduction, anti oxidant and lipid reduction properties [3] Numerous bioactive substances like Polypeptides, polysaccharides, alfa-glucans, pleuran, beta-glucan, lovastatin, natural statins, phenolic compounds and tannins have been estimated. The Indian strain *of P. djamor* (Arka-OM-1) has shown excellent iron chelating activity (78.5%) and excellent H_2O_2 scavenging activity [4].*P. cystidiosus* showed a good radical scavenging activity, stronger superoxide scavenging power and excellent reducing power[4].

Termitomyces species: India has a rich diversity of *Termitomyces* species which are relished in many of the regions. *T. medius, T. microcarpus, T. fuliginosus, T.eurhizus* and *T. letestui* h ave been documented. *Termitomyces spp.* was known by different local names as 'Beru Anabae' in Shimoga (Mushroom with root), 'Alande Kumi' in Kodagu and 'Kanda almi' in Konkani. **Regions of documentation from India:** Western Ghats, A&N islands, Sikkim, Gujarat and Rajasthan and Madhya Pradesh.

Medicinal properties: The species of *Termitomyces* have long been used for food and medicine in African countries. *T. microcarpus* is used to treat gonorrhea and for reducing cholesterol. In China it has been used to cure hemorrhoids, strengthen stomach and preventing intestinal carcinoma and as anti-oxidant. Bioactive constituents like cerebrosides Termitomycesphins have been isolated [5]

Clitocybe species: Indian forests have a rich diversity of both edible and non edible *Clitocybe species*. The edible species *C. maxima* is a relished mushroom in Gujarat where it is called as 'Vasrot' as it occurs in Bamboo groves. It is also relished in Madhya Pradesh where it is called as 'Baans piari' (Bamboo mushroom). The species has also been documented from Rajasthan. This species has been conserved at IIHR and experimental domestication has been successful.

Regions of documentation from India: Western Ghats, A&N islands, Sikkim, Gujarat and Rajasthan and Madhya Pradesh.

Medicinal properties: A laccase isolated from *C. maxima* showed antiproliferative activity against tumor cells and lowered the activity of HIV-1[6]

Coprinus disseminatus: Although not eaten in India, this species is cultivated in China and consumed as an important edible mushroom.

Regions of documentation from India: Western Ghats, A&N islands, Sikkim,

Medicinal properties: Mycelial extracts have shown to inhibit proliferation and induce apoptosis in human cervical carcinoma cells by activation of caspase, a key protein involved in the regulation of apoptosis[7]. Polysaccharides extracted from the mycelial culture and administered intraperitoneally into mice at a dosage of 300 mg/kg inhibited the growth of Sarcoma 180 and Ehrlich solid tumor by 100 and 90%, respectively[8].

Coprinus atramentarius: Commonly called as Ink cap mushroom, it is one of the common species seen on wood and ground. In China, this species is regarded as edible but is often grouped under toxic mushrooms due to its reactions of nausea and palpitations, when taken in conjunction with alcohol.

Regions of documentation from India: Western Ghats, A&N islands, Sikkim

Medicinal properties: Polysaccharides extracted from the mycelial culture of *C. atramentarius* and administered intraperitoneally into white mice at a dosage of 300 mg/kg inhibited the growth of Sarcoma 180 and Ehrlich solid cancers by 100%. [8]

***Tricholoma* species:** *Tricholoma* species are mycorrhizal species which have excellent edible mushrooms (*T. matsutakae*) as well as non edible species too. In India, the yellow caped (*T. sulphureum*) and violet caped (*T. nuda, Lepista nuda*) *Tricholoma* species have been documented. *T. nuda* is regarded as an excellent edible mushroom. *T. sulphureum* is a medicinal mushroom.

Regions of documentation from India: Western Ghats, Rajasthan, Sikkim and Madhya Pradesh.

Medicinal properties of *T. sulphureus*: Antitumor properties against Sarcoma 180 and Ehrlich solid cancers was reported by Ohtsuka *et al.* [8] A dosage of 300 mg/kg inhibited the growth by 90 and 80% in white mice.

***Calocybe species:* Calocybe** species are tropical edible species reported from Asia and Africa. The first indigenous mushroom to be commercialized by IIHR was *C. indica* (P & C) which was collected from the forests of Bengal. Another species of *Calocybe* was collected from the forests of Junagad in Gujarat which has been conserved at IIHR. The species is presently being studied for commercialization. *Calocybe gambosa* is the species relished in Africa and other countries.

Regions of documentation from India: West Bengal, Gujarat

Medicinal properties: *C. gambosa* shows antibacterial properties towards *Bacillus subtilis* and *Escherichia coli* [9]. *Calocybe indica* was found to contain good profile of carbohydrates with higher content of acid detergent fiber, neutral detergent fiber, hemicellulose and pectin. The functional efficacy of mushroom fiber was found equivalent to soluble fiber sample[10] . The authors also reported the lowering of lipidemic and glycemic indices by this mushroom. Brachvogel[11] also reported the the ability of *C. gambosa* to reduce blood sugar levels,

Lactarius piperatus: This species is commonly called as peppery milky cap as it secretes peppery latex.

Regions of documentation from India: Ravangla forest area of Sikkim

Medicinal properties: A new amino acid has been isolated from the peppery milkcap, named 2(S),3'(S)-1-(3-amino-3-carboxypropyl) -5-oxo-2-pyrrolidinecarboxylic acid. Four new marasmane sesquiterpenoids, lactapiperanols A-D, were isolated from the fruit bodies of *Lactarius piperatus* along with two known compounds lactarorufin and furosardonin[12]. Wang *et al.*[13] reported the occurrence of the following novel sesquiterpenes from the ethanol extract of *L.piperatus.*

- 7,8,13-trihydroxy-5,13-marasmanolide
- isoplorantinone
- 4,8,14-trihydroxyilludala-2,6,8-triene
- 8-hydroxy-8,9-secolactara-1,6-dien-5,13-olide

A natural liquid rubber has been found to be a small constituent of *L. piperatus* latex, composed largely of repeating *cis*-isoprene units[14] Jayko et al[15] reported the antitumor properties by hot water extract of *L. piperatus* which inhibited Lewis pulmonary adenoma in mice, with an inhibition rate of 80% against Sarcoma 180, and 70% against Ehrlich carcinoma. Antibacterial activity against *Escherichia coli, Proteus vulgaris,* and *Mycobacterium smegmatis was reported by Dulgar et al.*[16]. *Antioxidant activity was reported by Barros et al.* [17]

Auricularia species: It is an excellent culinary medicinal mushroom. Two species viz. *A. auricula,* and *A. polytricha* have been documented from India. The species have been conserved at IIHR.

Regions of documentation from India: Western Ghats, A&N and Sikkim

Medicinal properties: It is used for constipation, in excessive uterine bleeding and bleeding hemorrhoids, abdominal and tooth pain.[18] The Bioactive constituents reported are heteropolysaccharide glucans[19] and acid heteroglycans.[20]

Tremella species: Commonly called as Jelly fungus, the Chinese relish this mushroom although in India its usage is not known. Three species have been recorded from India. The white jelly fungus or the silver ear mushroom (*T. fuciformis*) from Western ghats and A&N islands, the yellow species commonly called as witches butter (*T .mesentrica)* and the brown species (*T. foliacea*) commonly called as Brown Witch's butter.

Regions of documentation from India: Western Ghats, A&N and Sikkim.

Medicinal properties: Cholesterol reduction, for arteriosclerosis, antherosclerosis and abnormal clotting, anti tumor, stimulates leukocyte activity[21]. Bioactive constituents reported are Polysaccharides, glycoproteins are effective in interferon production, increase splenocyte interleukin-2, promote phagocytosis.[18] *T. foliacea* shows antitumor effects[8], antibacterial activity.[22]

Trametes **species:** India has a great diversity of *Trametes* species. found in Western Ghats, A&N and Sikkim. The highest diversity was found in Western ghats.

Regions of documentation from India: Western Ghats, A&N islands, Rajasthan, Gujarat,Madhya Pradesh and Sikkim. The greatest diversity however was documented from Western ghats (Shimoga, Kodagu and Dakshin Kanada districts).

Medicinal properties: In China it has been used for the cure of upper respiratory, urinary and digestive tract, curative for liver ailments including hepatitis B, anti tumor, anti viral (HIV), immunostimulant,[23] increases interferon production in cancer patients.[24] One protein bound polysaccharide PSK and two polyoxygenated ergosterol derivatives have been isolated.

Schizophyllum **species:** Commonly known as split gill, this species is an important edible mushroom used in dry form in Manipur. The species has been conserved and domesticated at IIHR. One species *S. Commune* has been documented.

Regions of documentation from India: A&N islands, Rajasthan,

Medicinal properties: In China it is used to cure leucorrhea.[25] It has also been used for general weakness and debility,[25] to cure gynecological diseases. [26] Polysaccharide schizophyllan shows anti tumor activity, increased cellular immunity, restores killer cell activity and shows anti bacterial activity against *pseudomonas aeruginosa, Staphylococcus aureus, Escherichia coli and Klebsiella pneumoniae*

Pycnoporus cinnabarina: Commonly called as Orange polypore; The species has been conserved and dom esticated at IIHR. It can also be used as ornamental mushroom.[27]

Regions of documentation from India: Western Ghats (Kodagu, Dakshin Kanada) and Rajasthan.

Medicinal properties: This species shows antibacterial properties against *B. subtilis*. It shows biological activity against a variety of bacterial strains, with maximal inhibitory effect for Gram-positive bacteria of the genus *Streptococcus*. *P. cinnabarinus* produces the phenoxazinone derivative, cinnabarinic acid, a red pigment that accumulates in fruit bodies as well as in liquid cultures. Laccase secreted by the fungus oxidizes the precursor 3-hydroxyanthranilic acid to cinnabarinic acid, a reaction that is necessary for the production of antibacterial compounds. The biological activity of concentrated *P. cinnabarinus* culture fluid was nearly identical with that of cinnabarinic acid, synthesized by purified laccase *in vitro*.[28] The liquid culture filtrate of *Pycnoporus cinnabarinus* also shows good antibacterial effects against the growth of the Gram-negative bacteria *Escherichia coli* and *Pseudomonas aeruginosa* as well as Gram-positive *Staphylococcus aureus*. The culture filtrate was also used against mycelial growth and mycelial weight of three plant pathogenic fungi *Botrytis cinerea, Colletotrichum gloeosporioides [Glomerella cingulata]* and *Colletotrichum miyabeanus*, showing good inhibitory effect.[29] Polysaccharides extracted from the mycelial culture of *P. cinnabarinus* and administered intraperitoneally into white mice at a dosage of 300 mg/kg inhibited the growth of Sarcoma 180 and Ehrlich solid cancers by 90%.[8]

Podaxis species: This is a desert species growing in sandy soils of Rajasthan. In the region of Jodhpur, it is used both in fresh and dry form before sporulation. *Podaxis pistillaris* culture has been conserved at IIHR.

Regions of documentation from India: Gujarat and Rajasthan.

Medicinal properties: *It is* reported to be used in China to treat inflammation[30]. Antimicrobial activities against *Pseudomonas aeruginosa* and *Proteus mirabilis* have been reported[31]. Additionally, antibacterial activity against *Staphylococcus aureus, Micrococcus flavus,Bacillus subtilis, Proteus mirabilis, Serratia marcescens* and *Escherichia coli* is attributed to the epicorazines[32]. A high value for total lanthanides was measured in this mushroom (75 mg/kg dry weight) [33].In the culture medium of *P. pistillaris* three epidithiodiketopiperazines were identified as epicorazines A, B and C [32]. These three molecules appear to be structurally identical except for their chirality.

***Phellorinia* species:** Two species of this genus have been documented viz. *Phellorinia herculeana* and *P. inquinans*. This species grows on sandy calcerous soil. It is a relished edible and medicinal mushroom of Rajasthan (Jodhpur region) where it is called as 'Doda' and Gujarat (Bhuj region). This mushroom is used in the young stage for vegetable purpose and on maturity, the spore mass is mixed with flour for cooking.

Regions of documentation from India: Gujarat and Rajasthan.

Medicinal properties: In Jodhpur region, it is locally used as cure for impotency.

Polyporus aurcularis: Commonly called as Spring polypore, it has beautiful hexagonal pore surface and fine hairy cap margin.

Regions of documentation from India: Western ghats, Sikkim and Gujarat.

Medicinal properties: It shows antibacterial activity against *Escherichia coli, Salmonella typhimurium, Staphylococcus aureus* and *Bacillus subtilis* [33]. Antitumor activity against Ehrlich solid cancers in white mice was reported by Ohtsuka *et al.*[8], An inhibition of 90 and 100% was reported.

***Scleroderma* species:** These are the first species to occur after rains and may be found on soil or under specific trees. Commonly called as Puff balls, in India many of these species are eaten and relished. In Shimoga, *S. citrinum* is common and is commonly called as 'Koole anabe'. In A&N islands A species of *Scleroderma* is eaten and commonly called as 'Gurjan phootu' based on its association with Gurjan tree (*Dipterocarpus turbinatus)*. In Madhya Pradesh and Chattisgarh regions it is commonly called as 'Ban aloo' (wild potato) and is a relished mushroom.

Regions of documentation from India: Western ghats (Shimoga), A&N islands and Madhya Pradesh

Medicinal properties: The bioactive compounds isolated from *S. citrinum are*

- 23ξ-hydroxylanosterol was isolated from the peridium[34,35]

- Lanostane-type triterpenoid, which shows significant **antiviral** activity against Herpes simplex type 1

Jew's ear mushroom (*Auricularia delicata*)

Puff ball (*Scleroderma citrinum*)

Daldinia concentrica

Pycnoporus cinnabarina

Reishi mushroom *(Ganoderma lucidum)*

Dead man's finger *(Xylaria polymorpha)*

Silver ear mushroom *(Tremella fuciformis)*

Split gill mushroom *(Schizophyllum commune)*

Turkey tail *(Trametes versicolor)*

Cymatoderma spp.

- methyl 4,4'-dimethoxyvulpinate and two derivatives-dibromo and acetate-showed **antibacterial** activity towards the pathogen *Mycobacterium tuberculosis*; the derivatives were also cytotoxic to the NCI-H187 cell line

- 4,4'-dimethoxyvulpinic acid [36]

Pulvinic acid dimers have also been isolated from *Scleroderma citrinum*, including norbadione A, badione A and sclerocitrin. The latter compound is a pigment that gives the fungus its yellow color, and is present in relatively high amounts-400 mg/kg fungus [37].

Cymatoderma **species:** This is a beautiful polypore of rare occurrence.

Regions of documentation from India: A&N islands (Rutland)

Medicinal properties: Polysaccharides extracted from the mycelial culture of *C. elegans* and administered intraperitoneally into white mice at a dosage of 300 mg/kg inhibited the growth of Sarcoma 180 and Ehrlich solid cancers by 80% and 70%, respectively[8].

Fistulina hepatica: Commonly called as Red polypore, it is a choice mushroom in Sikkim.

Regions of documentation from India: Sikkim. It is called as 'Lal chiu' by the locals and relished by them.

Medicinal properties: *F. hepatica* has been investigated for its capacity to act as a free-radical scavenging agent. Good radical-scavenging activity was noted against DPPH, the superoxide radical and the hydroxyl radical, while only a weak protective effect was seen against hypochlorous acid[38]. Some Italian studies by Coletto have revealed that *F. hepatica* has potent antibacterial activity against *Escherichia coli, Staphylococcus aureus, Bacillus subtilis* [39,40], and *Klebsiella pneumoniae* 41.Also, the tetrayne-tetraol mentioned above has modest antibacterial activity, comparable with that of cephalosporin C against *Staphylococcus aureus* and *Salmomlla typhi*, when tested by the hole-plate method[42]. The growth medium affects the nematicidal activity of *F. hepatica* (and various other fungi). When grown in Czapek broth, filtrates of the beefsteak fungus were pathogenic to the wood nematode,*Bursaphelenchus xylophilus* grown *in vitro*. However, grown in potato dextrose broth, filtrates of the same fungus were not pathogenic[43]. Polysaccharides extracted from the mycelial culture of *F. hepatica* and administered intraperitoneally into white mice at a dosage of 300 mg/kg inhibited the growth of Sarcoma 180 and Ehrlich solid cancers by 80% and 90%, respectively[8].

Geastrum triplex: Commonly called as Earth star due to breaking up of mesoperidium in star shaped petals.

Regions of documentation from India: Western ghats, Sikkim.

Medicinal properties: The fruit bodies of *Geastrum triplex* Jungh contain the sterols ergosta-4,6,8,(14),22-tetraen-3-one,5,6-dihydroergosterol, ergosterol, peroxyergosterol, as well as myristic, palmitic, stearic, oleic, linoleic, and linolenic acids [44].

Hydnum repandum: Commonly called as tooth fungi due to teeth like projections of the pore region.

Regions of documentation from India: Sikkim.

Medicinal properties: An extract of the culture mycelia showed 70% inhibition againstSarcoma 180 solid cancer in mice, while extracts from the fruit bodies showed 90% inhibition against both Sarcoma 180 and Ehrlich solid cancer in mice[8] (Ohtsuka_et al., 1973). Furthermore, the compound repandiol described above showed potent cytotoxic activity against a variety of tumor cell types, especially colon adenocarcima cells, for which its IC_{50} is 0.30[45]. In a study of antimicrobial activity using the disk-diffusion method, it was shown that a chloroform extract of the hedgehog mushroom had mild antibiotic activity against *Enterobacter aerogenes, Staphylococcus aureus, Staphylococcus epidermidis* and *Bacillus subtilis,* while the ethanol extract had mild activity against only *Bacillus subtilis* [46].

Psuedohydnum gelatinosum: Commonly called as toothed jelly fungus, false hedgehog fungus is of rare occurrence.

Medicinal properties: This is the only species which contains Lectins with anti-A serologic specificity with human and Rabbit blood cells[47]. Polysaccharides extracted from the mycelial culture of *P. gelatinosum*and administered intraperitoneally into white mice at a dosage of 300 mg/kg inhibited the growth of Sarcoma 180 and Ehrlich solid cancers by 90%[8].

Ganoderma **species***:* One of the most utilized and often referred to as king of medicinal mushrooms. It has been documented from all the regions under study. *G. lucidum* was more prevalent in Rajasthan as compared to other species like *G. aplanatum and G. tsugae.* India has a rich diversity of Ganoderma species. Forty five *Ganoderma* isolates were collected by IIHR from different parts of the country (Karnataka-23, Rajasthan-8, Andaman-6, H.P-1 and Gujarat-7). The variability in morphological and cultural characters of these isolates was studied in detail. The isolates in same locality also showed high variability in their characters. Forty three isolates were successfully cultivated on sawdust medium[48,49]. The isolates exhibited high variability in terms of spawn run, days required for primordial initiation, morphological characters of fruiting body yield and dry recovery. The fruit body colour varied with isolates. The color shown by the fruiting bodies include yellowish, orange, orangish red, orangish brown, chocolate brown, brown, dark brown etc. The shape of the fruiting body varied from irregular, conical to typical kidney shape. An isolate G-33 from Andaman produced antlers only. The investigation on various aspects of *Ganoderma* conducted at IIHR clearly indicates the high variability of the species existing in the country and its enormous potential for commercialization. Successful cultivation of the precious mushroom on sawdust/ wheat straw was reported by various other workers in India[50,51].

Regions of documentation from India: Western ghats, Rajasthan, Gujarat, A&N islands

Medicinal properties: *Ganoderma* species have been used as medicinal mushrooms in China and Southeast Asia.[52]. *Ganoderma* neutriceuticals are used to treat patients suffering from different illnesses including cardiovascular problems, cancer, leukemia, leucopoenia, hepatitis, nephritis, gastritis, insomnia, cholesterol lowering, asthma and bronchitis. Because of its perceived health benefits, fruiting body has gained wide popularity recently as a dietary supplement, not only in China and Japan but also in North America and other parts of the world[53].

Dictyophora species: Commonly called as veiled lady mushroom, it is one of the most delicate of culinary/medicinal mushrooms. Two species *viz. D. indusiata* and *D. duplicata* have been documented. It is a prized edible mushroom in China where it is served in importants banquets.

Regions of documentation from India: Western ghats, A&N islands, Sikkim.

Medicinal properties: It posses good antioxidant and antimicrobial properties[54].

Hericium **species:** Commonly called as monkey head mushroom, one species *H. erinaceus* has been documented from Rajasthan and Sikkim. The species has been conserved and domesticated at IIHR.

Regions of documentation from India: Sikkim, Rajasthan.

Medicinal properties: The extract of this species is approved by the Chinese ministry of health for use in chronic superficial gastritis (CSG). It stimulates immune system, has antioxidant and antitumor properties too. The species is also being studied for its ability to stimulate nerve cells[55].

Clavaria vermicularis: **Commonly called as Fairy fingers, this beautiful mushroom was documented from Western Ghats and Sikkim. Medicinal properties:** An extract of the fruit bodies of *C. vermicularis* inhibited the growth of Sarcoma 180 and Ehrlich solid cancers in mice by 90% and 80%, respectively[8].

Xylaria polymorpha: Commonly called as Dead man's fingers.

Regions of documentation from India: Western ghats, Rajasthan, Gujarat, Madhya Pradesh.

Medicinal properties: In Indian traditional medicine (Ayurvedic medicine), the powdered fruit bodies, mixed in equal proportion with sugar, is used to promote lactation after birth[56]. 2-Hexylidene-3-methylsuccinic acid, aka piliformic acid, is the major metabolite produced by *X. polymorpha*[57].This compound showed moderate cytotoxicity against KB and BC-1 cell lines [58]. Dead man's fingers was shown to contain about 6% mannitol (dry weight), a sugar used as a diuretic agent[59]. Two new polypropionates designated as xylarinic acids A (4,6,8-trimethyl-2,4-decadienoic acid) and B (2,4,6-trimethyl- 2-octenoic acid) were isolated from *X. polymorpha* fruiting bodies. Both compounds displayed significant antifungal activity against the pathogenic plant fungi *Pythium ultinum, Magnaporthe grisea, Aspergillus niger, Alternaria panax,* and *Fusarium oxysporium,* but they did not show any antibacterial nor cytotoxic effects[60].

Daldinia: Commonly called as cramp balls or King Alfred's cake, it is an inedible species found growing on wood.

Regions of documentation from India: Western Ghats, A&N islands, Madhya Pradesh, Rajasthan, Gujarat, Sikkim.

Medicinal properties: A new derivative of benzofuran lactone, named concentricolide (1), was isolated from the fruiting bodies of the ascomycete *Daldinia concentrica*. This compound showed *in vitro* anti HIV activity.

Sarcoscypha and Scutellinia **species**: Commonly called as saucer or cup fungi and eyelash cup. *Sarcoscypha occidentalis* and *Scutellinia scutellata* are the species documented from the regions explored.

Regions of documentation from India: Western ghats, Sikkim, A&N islands.

Medicinal properties: *Sarcoscypha coccinia* shows good antioxidant properties[61].

Conclusion:

India has a great diversity of both edible and medicinal mushrooms, many of which are yet to be researched for their neutraceutical properties. There is a need to undertake collaborative research between mycologists, chemists, pharmacologists and medical institutions to harness the complete potential of this mycological power nature has bestowed. The collection and conservation of mushrooms is a specialized task which is region specific. Due to its very short shelf life, a non regular occurrence the conservation studies have to be done locally of small regions so that an effective and viable germplasm collection with relevant documentation can be conserved for its effective utilization in the future. There is an urgent need to form a National data bank of the important edible and medicinal mushrooms so that further research on the bioactive constituents of various mushrooms can be undertaken.

References

1. Hawksworth, D.L. The fungal dimension of biodiversity: magnitude, significance, and conservation. *Mycol Res*, 1991, **95**: 641-655.

2. Arora, D. Mushrooms demystified 1986, Ten speed press.

3. Ajith, T.A., and Janardhanan, K.K., Indian Medicinal mushrooms as a source of Antioxidant and antitumor agens, *J. Clin Biochem Nutr.* 2007, 40,3, 157–162.

4. Dandamudi Rajesh Babu., Meera Pandey and G, Nageswara Rao., Antioxidant and electrochemical properties of cultivated *Pleurotus spp.* and their sporeless/ low sporing mutants. *J Food Sci Technol*, 2012, s13197-012.

5. Yuvan Q., Kaiyue S., Lijuan G., Youji, S., Horokaju, K., Makoto, O., and Jianhua, Q. Termitomycesphins G and H, Additional cerebrosides from edible Chinese mushroom *Termitomyces albuminosus. Biosci. Biotechnol. Biochem*, 2012, 76, **4**, 110918-1-3.

6. Zhang, G. Q., Wang, Y. F., Zhang, X. Q., Ng, T. B., and Wang, H. X., Purification and characterization of a novel laccase from the edible mushroom Clitocybe maxima, *Process Biochem*, 2010,**5**, 627–633.

7. Han,B.Y., Toyomasu,T, and Shinozawa,T.Induction of apoptosis by *Coprinus disseminatus* mycelial culture broth extract in human cervical carcinoma cells. *Cell Structure and Function* 24,**4**, 209-15.

8. Ohtsuka, S., Ueno, S., Yoshikumi, C., Hirose, F., Ohmura, Y., Wada, T., Fujii, T. and Takahashi,E., Polysaccharides having an anticarcinogenic effect and a

method of producing them from species of Basidiomycetes. UK Patent 1331513., 26 September 1973.

9. Keller, C., Maillard, M., Keller, J. and Hostettmann, K., Screening of European fungi for antibacterial, antifungal, larvicidal, molluscicidal, antioxidant and free-radical scavenging activities and subsequent isolation of bioactive compounds. *Pharmaceutical Biol.*, 2002, **40**,7, 518-25.

10. Tapasya, A and Rashmi, K., Determination of in vitro functional efficacy of fibre isolate from mushrooms. *J Dairying Foods Home Scie*, In Press., 2012.

11. Brachvogel, R., Reduction of blood sugar by *Calocybe gambosa* Fr. Donk. Zeitschrift für Mykologie., German., 1986, **52**, 445.

12. Yaoita, Y., Machida, K. and Kikuchi, M., Studies on the constituents of mushrooms, part V - Structures of new marasmane sesquiterpenoids from *Lactarius piperatus* (SCOP.: Fr.) S. F. Gray. *Chem Pharmaceutical Bulletin.* 1999b, **47**, 894-6.

13. Wang,Y., Yang, S.P., Yue, J.M., Chow, S. and Kitching, W., Novel sesquiterpenes from the fungus *Lactarius piperatus*. *Helvetica Chimica Acta.*, 2003, **86**, 2424-33.

14. Tanaka, Y., Kawahara, S., Eng, A.H., Takei, A. and Ohya, N., Structure of *cis*-polyisoprene from *Lactarius* mushrooms. *Acta Biochim Pol.*, 1994, **41**, 303-9.

15. Jayko, L.G., Baker, T.I., Stubblefield, R.D. and Anderson, R.F., Nutrition and metabolic products of *Lactarius* species. *Canadian J Microbiol*, 1974, **8**, 361–371.

16. Dulger, B., Yilmaz, F. and Gucin, F., Antimicrobial activity of some *Lactarius* species. Pharm Biol., 2002, **40**, 304-6.

17. Barros, L., Baptista, P. and Estevinho, L.M., Ferreira ICFR. Effect of fruiting body maturity stage on chemical composition and antimicrobial activity of *Lactarius* sp. mushrooms.*J Agric Food Chem.*,2007, **55**,8766-8771.

18. Hobbs, C., Medicinal Mushrooms, Botanica Press, Santa Cruz, l995.

19. Misaki, A. Studies on the interrelation of structure and antitumor effects ofpolysaccharides:: Antitumor action of periodate-modified, branched (1-3-beta-Glucan) ofAuricularia auricula-judae, and other polysaccharides containing (1-3)-Glycosidic Kikurages. *Carbohyd. Res.*1981, **92**:115-129.

20. Ukai, S., Kiho T., Hara, C., Kuruma, I. and Tanaka, Y.. Polysaccharides in fungi. Anti-inflammatory effect of the polysaccharides from the fruit bodies of several fungi, *Journal of Pharmacobiology* 1983,**6**,983–90.

21. Hu, B. and But, P. Chinese materia medica for radiation protection. *Abstracts of Chinese Medicines*, 1987,**1**,475-490.

22. McCormack, P.J., Wildman, H.G. and Jeffries, P., Production of antibacterial compounds by phylloplane-inhabiting yeasts and yeastlike fungi. *Appl Environ Microbiol.*, 1994, **60**,927-31.

23. Tsukagoshi S, Hashimoto Y, Fujii G, Kobayashi H, Nomoto K and Orita K, Krestin (PSK).*Cancer Treatment Reviews*, 1984, **11**, 131–55.

24. Ebina, T., Kohya, H., Inoue, H., Antitumor effect of PSK (1) Interferon inducing activity and intratumoral administration. *Gon to Kagaku Ryoho* 1987a, **14,**1841-1846.

25. Ying, J., Mao, X., Ma, Q., Zong, Y. and Wen, H. Icones of medicinal fungi from China. Science Press, Beijing, China, 1987. pp 575.

26. Yang, Q., and Jong, S., Medical mushrooms in China, in Proceedings of the Twelfth International Conference on the Science and Cultivation of Edible Fungi, 1989XII:631-643.

27. Pandey, M. and Veena S.S.. Medicinally important mushrooms, In: Lifestyle Horticulture. (T. Janakiram, K.V. Prasad, K.P. Singh & Kishem Swaroop eds). 2010,**48**, 118-124

28. Eggert, C., Laccase-catalyzed formation of cinnabarinic acid is responsible for antibacterial activity of *Pycnoporus cinnabarinus*. *Microbiol. Res.*, 1997, **152**, 315-8.

29. Imtiaj, A. and TaeSoo, L., Screening of antibacterial and antifungal activities from Korean wild mushrooms. *World J Agric Sci.*, 2007, **3**, 316-21.

30. Mao, X.L., The Macrofungi in China.Zhengzou: *Henan Science and Technology Press*, 2000. (Chinese)

31. Panwar, C. H., and Purohit, D. K. Antimicrobial activities of *Podaxis pistillaris* and *Phellorinia inquinans*against *Pseudomonas aeruginosa* and *Proteus mirabilis*. *Mushroom Res.* 2002, **11,**43–4.

32. Al-Fatimi,M.A., Julich,W.D., Jansen,R., and Lindequist, U., Bioactive components of the traditionally used mushroom *Podaxis pistillaris*. Evid Based Complement *Alternat Med.* 2006 3,1,87-92.

33. Stijve, T., Andrey, D., Lucchini, G. F. and Goessler, W., Simultaneous uptake of rare elements, aluminium, iron, and calcium by various macromycetes. *Aust Mycol.* 2001; **20**:92–8.

34. Entwistl, N., Pratt, A.D., 23ξ-hydroxy-lanosterol a new triterpene fungal metabolite of basidiomycete *Scleroderma aurantium* Pers. Tetrahedron., 1968, **24**, 3949-53.

35. Entwistl, N., Pratt, A.D., Determination of absolute configuration at C23 in 23-hydroxylanosterol - a triterpene fungal metabolite of basidiomycete *Scleroderma aurantium* Pers and its C23 epimer.Tetrahedron., 1969, **25**, 1449-51.

36. Kanokmedhakul, S., Kanokmedhakul, K., Prajuabsuk, T., Soytong, K., Kongsaeree, P., Suksamrarn, A., A bioactive triterpenoid and vulpinic acid derivatives from the mushroom *Scleroderma citrinum*. Planta Med.. 2003, **69**, 568-71.

37. Winner, M., Gimenez, A., Schmidt, H., Sontag, B., Steffan, B., Steglich, W., Unusual pulvinic acid dimers from the common fungi *Scleroderma citrinum* (common earthball) and *Chalciporus piperatus* (peppery bolete). Angew Chem Int Ed Engl., 2004, **43**, 1883-6.

38. Ribeiro, B., Valentão, P., Baptista, P., Seabra, R.M., Andrade, P.B., Phenolic compounds, organic acids profiles and antioxidative properties of beefsteak fungus *Fistulina hepatica*. Food Chem Toxicol., 2007, **45**, 1805-13.

39. Coletto, M.A.B., Basidiomycetes in relation to antibiosis. II. Antibiotic activity of mycelia and culture liquids. G,Batteriol Virol Immunol. Italian., 1981, **74**, 7-12,267-74.

40. Coletto, M.A.B., Antibiotic activity in basidiomycetes III. Antibacterial activity of mycelia and culture filtrates. Allionia, Turin., 1988, **28**,165-70.

41. Coletto, M.A.B., Antibiotic activity in basidiomycetes. VI. Antibiotic activity of mycelia and cultural filtrates of thirty three new strains. Allionia, Turin., 1992, **31**,87-90.

42. Farrell, I.W., Keeping, J.W., Pellatt, M.G., Thaller, V., Natural acetylenes Part 41. Polyacetylenes from fungal fruiting bodies. J Chem Soc Perk Trans I., 1973, **22**, 2642-3.

43. Dong, J.Y., Li,X, P., Li, L., Li, G.H., Liu, Y.J., Zhang, K.Q., Preliminary results on nematicidal activity from culture filtrates of Basidiomycetes against the pine wood nematode, *Bursaphelenchus xylophilus* (Aphelenchoididae). Ann Microbiol. 2006, **56**,163-6.

44. Torpoco, V., Garbarino, J.A., Studies on Chilean fungi. I. Metabolites from *Geastrum triplex* Jungh. *Boletin De La Sociedad Chilena De Quimica.*, 1998, **43**, 227-9.

45. Takahashi, A., Endo, T., Nozoe, S., Repandiol., A new cytotoxic diepoxide from the mushrooms *Hydnum repandum* and *H. repandum* varalbum. Chem Pharm Bull., 1992, **40**, 3181-4.

46. Yamach, M., Bilgili, F., Antimicrobial activities of fruit bodies and/or mycelial cultures of some mushroom isolates. Pharm Biol., 2006, **44**, 660-7.

47. Pemberton, R.T. Agglutinins (lectins) from some British higher fungi. *Mycol Res.* 1994 98:277-90 Part 3.

48. Veena, S. S., and Meera Pandey, Evaluation of the locally available substrates for the cultivation of indigenous *Ganoderma* isolates. *J. Mycol.Pl. Pathol*, 2006, **36**, 434 – 438

49. Veena. S.S and Meera Pandey, Paddy Straw as a Substrate for the Cultivation of Lingzhi or Reishi Medicinal mushroom, *Ganoderma lucidum* (W. Curt.:Fr.) P.Karst.in India. *International Journal of Medicinal Mushrooms,2011*, **13**, 397-400.

50. Rai R. D., Successful cultivation of the medicinal mushroom Reishi, *Ganoderma lucidum* in *Indian.Mushr Res.* 2012, **2**, 87-91.

51. Mishra, K. K. and Singh, R.P., Exploitation of indigenous *Ganoderma lucidum* for yield on different substrates. *J. Mycol.Pl. Pathol*, 2006, **36**,130 –133.

52. Chen, A,W., Natural – log cultivation of the medicinal mushroom *Ganoderma lucidum* (Reishi). *Mush world. com.* 2004-01-09.

53. Chang, S.T., *Ganoderma lucidum* (A leader of medicinal mushrooms): A prominent source for the health care market in the 21 st century. *Mush world. com.* 2005-02-02

54. Oyetayo, V.O., Dong, C.-H. and Yao, Y.J., Antioxidant and Antimicrobial Properties of Aqueous Extract from *Dictyophora indusiata, The Open Mycol J.*, 2009; **3**,20-26

55. Mori K, Obara Y, Hirota M, Azumi Y, Kinugasa S, Inatomi S, Nakahata N. Nerve growth factor-inducing activity of Hericium erinaceus in 1321N1 human astrocytoma cells. *Biol Pharm Bull.* 2008, **31**, 727-32.

56. Rai, M., Mandal, S.C., Acharya, K., A proximate composition of *Fistulina hepatica*, an endemic mushroom of the Darjeeling hills. *J Hill Res.*, 2007, **20**,39-41.

57. Anderson, J.R., Edwards, R, L., Whalley, A.J.S., Metabolites of the higher fungi. 22. 2-butyl-3-methylsuccinic acid and 2-hexylidene-3-methylsuccinic acid from Xylariaceous fungi. *Chem Soc Perk Trans,* I. 1985, **7**,1481-6.

58. Chinworrungsee, M., Kittakoop, P., Isaka, M., Rungrod, A., Tanticharoen, M. and Thebtaranonth Y., Antimalarial halorosellinic acid from the marine fungus *Halorosellinia oceanica.Bioorg Med Chem Lett.* 2001 **11**,1965-9.

59. Snatzke, G., and Wolff, H.P., Mannitol from *Xylaria polymorpha.* Zeitschrift für Mykologie. 1987, **53**,137-8.

60. Jang, Y.W., Lee, I. K., Kim, Y. S., Lee, S., Lee, H. J., Yu, S, H., and Yun, B. S. Xylarinic acids A and B, new antifungal polypropionates from the fruiting body of *Xylaria polymorpha.J Antibiotics.* 2007, **60**,696-9.

61. Wani, A. H., Boda, R. H., Taskeen-un-Nisa and Latif A. Potential antioxidant activity of some mushrooms growing in Kashmir Valley *Mycopath*, 2010, **8**, 71-75.

16

Evaluation of the safety of proprietary products of *Ganoderma lucidum* (Fr.) P. Karst : a preclinical study

Ravindran K. Veena[1], T.A. Ajith[2] and K.K. Janardhanan[1*]

[1]Department of Microbiology, Amala Cancer Research Centre, [2]Department of Biochemistry, Amala Institute of Medical Sciences, Amala Nagar, Thrissur – 680 555, Kerala, India. *E-mail: kkjanardhanan@yahoo.com

Reishi or Lingzhi – *Ganoderma lucidum* based products have attracted great attention during the last two decades, not only in South East Asia but also in North America and Europe. Acute and chronic toxicity of powdered fruiting bodies as well as mycelium of two proprietary products of *G. lucidum*, Reishi Gano (RG) and Ganocelium (GL) were evaluated using Swiss albino female mice and Swiss albino female rat, respectively. For testing acute toxicity, single dose of 500 and 2500 mg/kg body weight were administered and animals were observed for 72 hrs for toxic symptoms. For testing chronic toxicity, the products suspended in water were administered (100 and 250 mg/kg body weight) once daily for 90 days. After the last dose of drug administration, animals were sacrificed and blood was collected. Serum transaminases (GOT and GPT) and alkaline phosphatase (ALP), albumin, total protein, urea and creatinine were determined in the serum samples. The biochemical parameters of the liver function test indicate no significant increase in the activity of SGPT, ALP and SGOT, or in the level of total protein and albumin in any of the treated group of animals compared to the normal group of animals. Similarly, renal function tests such as serum urea and creatinine levels were not elevated in any of the treated group of animals. Lipid profile and glucose levels were also unaltered in any of the treated group of animals. The results of this study concluded that proprietary products of *G. lucidum*, Reishi Gano (RG) and Ganocelium (GL) did not show any acute or chronic toxicity when administered on short term or long term basis. The results suggest that *Ganoderma* proprietary products tested in this study are safe for consumption.

Introduction

The importance of fungi in modern medicine was recognized long back with the discovery of penicillin. Some of the macrofungi have been used from time immemorial in folk medicine. Many pharmaceutical substances with potent and unique health-enhancing properties have been isolated from medicinal mushrooms[1]. Recently demonstrated medicinal properties include antitumor, inmmunomodulating, antioxidant, cardiovascular, antiviral, antibacterial, antiparasitic, antifungal, hepatoprotective and antidiabetic. Most of the mushroom products are dried powders and extracts from naturally growing or commercially cultivated mushroom fruiting bodies or biomass of mycelium grown in a solid state or in a submerged culture. It is generally believed that many mushrooms can be safe because of their long history of usage. However, the scientific data on the safety of many of the commercial and proprietary mushroom preparations are lacking.

According to Chinese tradition, *G. lucidum*, popularly known as Reishi or Lingzhi is a miraculous herb, bringing good health, longevity and even immortality. Products derived from the powdered fruiting bodies of Reishi have been demonstrated to possess preventive and curative effects on diseases such as heart diseases, hypertension, hepatitis, diabetes, neurasthenia, tumor and cancer. *Ganoderma* based products have become popular not only in South East Asia but also in North America and Europe. Although *Ganoderma* based products are considered safe, available information on the toxicity of these products is inadequate. Clinical toxicology can be defined as the study of the clinically significant changes caused by xenobiotic and or therapeutic exposure, which are adverse in nature for the patient. Acute toxicity is usually defined as the adverse changes occurring immediately or at a short time following the administration of a single exposure of substance [2].

Toxicity evaluation of the powder or extract when administered for long duration has not been reported till now. Hence, present study was undertaken to evaluate the acute and chronic toxicity of two proprietary products of *G. lucidum*, Reishi Gano (RG) and Ganocelium (GL) and the finding are reported in this paper. These products are gift samples from Daxen Agritech India Pvt. Ltd.

Experimental

Evaluation of toxicity on animal models

The products, Reishi Gano (capsules of micro fine powder of fruiting bodies) and Ganocelium (capsules of micro fine powder of mycelium) were used in this study. The capsules were suspended in sterilized water for administration.

Animals

Female Swiss albino rat weighing 250 ± 2 g were used for the chronic toxicity study and Swiss albino mice (25 ± 2g) were employed for acute toxicity study. Animals were initially purchased from Small Animal Breeding Centre, Kerala Agricultural University, Mannuthy, Thrissur, Kerala, India and were kept for a week under environmentally controlled conditions with free access to standard food and water *ad libitum*. The experiments were carried out according to the guidelines of Committee for the Purpose of Control and Supervision of Experiments on Animals (CPCSEA), Government of India and approved by the Institutional Animal Ethics Committee, Amala Cancer Research Centre, Amala Nagar, Thrissur, Kerala, India. The testing was done following Organization of Economic Co-operatin and Development (OECD) guideline for testing of chemicals.

Acute toxicity study

Swiss albino mice (25 ± 2 g) were used for the study. Animals were divided into five groups of six animals in each group. The drug was administered orally as single dose. Group I was administered with distilled water and kept as normal. Group II and Group III were treated with Reishi Gano, 500 and 2500 mg/kg body weight and Group IV and V with Ganocelium at doses of 500 and 2500 mg/kg body weight respectively. The animals were observed for toxic symptoms and mortality for 72 hrs.

Chronic toxicity

Swiss albino rat were used for the study. Animals were divided into five groups of six animals each. The drug was administered orally as follows. Group I administered with distilled water was kept as normal. Group II and Group III were treated with Reishi Gano (100 and 250 mg/kg body weight) where as Group IV and V were administered with Ganocelium (100 and 250 mg/kg body weight), once daily for 90 days. The body weight changes were recorded weekly. One day after the last dose of Reishi Gano and Ganocelium administration animals were sacrificed after an overnight fasting. The blood was taken out by heart puncture. Serum transaminases (SGOT and SGPT) and alkaline phosphatase (ALP), albumin, total protein, urea and creatinine, lipid profile such as total cholesterol, triacyl glycerol, HDL and LDL cholesterol were evaluated in the serum sample using commercially available diagnostic kits using a double beam spectrophotometer.

In acute toxicity study Reishi Gano and Ganocelium did not show any toxic symptoms. The animals were healthy, there was neither mortality nor toxic symptoms up to the dose of 2500 mg//kg body weight 72 hrs after administration.

In chronic toxicity studies, treatment with Reishi Gano and Ganocelium did not produce any toxicity and also did not show significant change in the body

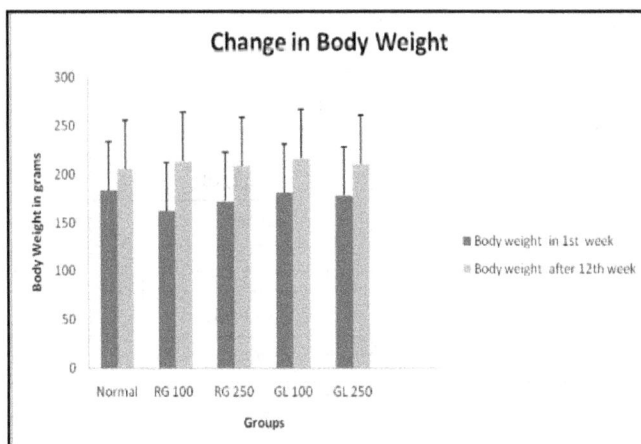

Fig. 1: Effect of Reishi Gano (RG) and Ganocelium (GL) on body wt.

weight when compared to the normal group of animals (Fig 1). Further, treatment for 90 days did not produce any significant changes in the liver function or renal function test parameters compared to the normal group of animals (Table 1). A slight increase in the mean value of SGOT observed in the Ganocelium (250 mg/kg body wt.) treated group of animals however, it was non significantly different from that of the normal group. The concentration of serum creatinine and urea are given in table 2. No change in the urea or creatinine levels were observed between the normal and the Reishi Gano and Ganocelium treated groups.

Similar result was also found among the lipid profile parameters. Triacyl glycerol level was found to be elevated in the GL (250 mg/kg) treated group. A significant increase was evidenced in the level of HDL cholesterol, in the entire Reishi Gano and Ganocelium treated group without any significant increase in the total cholesterol level (Table 4).

Table 1: Effect of *Ganoderma*, Reishi Gano (RG) and Ganocelium (GL), on the serum enzyme activity

Groups	Treatment (mg/kg)	SGPT (IU/L)	SGOT (IU/L)	Total protein (g/dl)	Albumin (g/dl)	ALP (IU/L)
Normal	Vehicle	29.6 ± 5.78	65.6 ± 12.3	7.4 ± 0.46	3.8 ± 0.52	39.9 ± 6.1
RG	100	17.7 ± 5.50	76.2 ± 5.6	7.8 ± 0.59	3.9 ± 0.39	31.2 ±9.1
	250	29.0 ± 8.83	69.8 ± 6.2	7.6 ± 0.65	3.7 ± 0.36	40.8 ± 6.8
GL	100	23.7 ± 2.40	65.3 ± 4.1	7.5 ± 0.05	3.6 ± 0.16	41.2 ± 5.9
	250	19.2 ± 7.70	70.5 ± 4.4	7.9 ± 0.38	3.7 ± 0.32	40.8 ± 4.8

Values are mean ± S.D (n= 6).

P > 0.05 (Bonferroni test) was non-significantly different from normal group.

Table 2: Effect of *Ganoderma*, Reishi Gano (RG) and Ganocelium (GL), on the renal function test parameters

Groups	Treatment (mg/kg)	Urea (mg/dl)	Creatinine (mg/dl)
Normal	Vehicle	40.8 ± 2.7	0.52 ± 0.08
RG	100	38.9 ± 3.7	0.43 ± 0.1
	250	41.0 ±1.9	0.56 ± 0.2
GL	100	37.6 ± 3.1	0.54 ± 0.2
	250	38.5 ± 2.0	0.64 ± 0.1

Values are mean ± S .D (n= 6).

P > 0.05 (Bonferroni) was non-significantly different from normal group

Table 3: Effect of *Ganoderma*, Reishi Gano (RG) and Ganocelium (GL), on the lipid profile

Groups	Treatments (mg/kg)	Triglycerides (mg/dl)	Total Cholesterol (mg/dl)	HDL (mg/dl)	LDL (mg/dl)
Normal	Vehicle	137.4 ± 16.3	106.7 ± 4.5	14.0 ± 0.1	58.2 ± 0.6
RG	100	136.1 ± 19.9	105.4 ± 1.2	22.3 ± 1.1*	58.2 ± 4.4
	250	161.9 ± 20.5	106.4 ± 3.2	23.7 ±1.1*	50.7 ± 3.1
GL	100	156.3 ± 15.4	111.5 ± 5.1	28.3 ± 2.8*	56.7 ± 4.2
	250	163.1 ± 17.0	108.3 ± 5.1	27.1 ± 1.8*	56.6 ± 5.4

Values are mean ± S .D (n= 6).

*$P < 0.05$ (Bonferroni test) significantly and $P>0.05$ was non significantly different from the normal group

Table 4: Effect of *Ganoderma*, Reishi Gano (RG) and Ganocelium (GL), on the serum glucose level

Groups	Treatments (mg/kg)	Glucose (mg/dl)
Normal	Vehicle	88.22 ± 5.50
RG	100	90.50 ± 4.86
	250	93.27 ± 5.09
GL	100	87.02 ± 1.16
	250	90.26 ± 3.80

Values are mean ± S .D (n= 6).

$P > 0.05$ (Bonferroni test) non significantly different from the normal group

Reishi- *G. lucidum* has been demonstrated to possess several pharmacological properties and therapeutic uses[3-5]. Contemporary Tradition Chinese Medicine uses products of cultivated *G. lucidum* for treatment of neurasthenia, debility from prolonged illness, insomnia, anorexia, dizziness, chronic hepatitis, hypercholesterolemia, coronary heart diseases, hypertension, altitude sickness and cancer [6.] Results of the study reveal that Reishi Gano and Ganocelium did not show any toxicity when administered for a long term. It was previously found that administration of *G. lucidum* up to 2500 mg/kg body wt. did not produce any lethal effect to animals and LD_{50} could not be determined. Serum transaminases (GOT and GPT) and ALP activities are good indices of liver damage. The biochemical parameters to evaluate the liver function test indicate no significant increase in the activity of SGPT, ALP and SGOT in any of the Reishi Gano and Ganocelium treated group of animals compared to the normal animals. A slight (statistically non-significant) increase in the SGOT activity in the Ganocelium (250 mg/kg) treated group of animals compared to normal was observed. However, no considerable changes in other parameters such as SGPT, total protein, albumin or globulin levels were found in any of the treated group of animals. This indicates that the observed small change in the liver function enzyme, SGOT does not seem to be sufficient to support the toxicity of Ganocelium. Further, Reishi Gano and Ganocelium did not induce any damage to kidneys as evident from the normal level of urea and creatinine.

Safety and tolerability of *G. lucidum* were tested in a randomized placebo-controlled human trail[7]. The reported findings indicated that compared to placebo group, no adverse effects were observed after the administration of the mushroom extract. Results of current study conclude that proprietary products of *G. lucidum*, Reishi Gano (RG) and Ganocelium (GL) manufactured by Daxen Agritech India Pvt. Ltd do not cause any significant toxicity when administered for a long period.

References

1. Wasser, S.P., Medicinal mushroom science: History, current status, future trends, and unsolved problems. *Int. J. Med. Mushr.* 2010,**12**,1-16.

2. Walum, E., Acute oral toxicity, *Environ. Health Perspect* .1998,**106**,497-503.

3. Jonn, W.M. and Gohel, M. D. I., Anticancer effects of *Ganoderma lucidum* : A review of scientific evidence. *Nutr. Cancer.* 2005,**53**,11-17.

4. Paterson, R.R,M., *Ganoderma* –A therapeutic biofactory. *Phytochemistry* 2006,**67**, 1985-2001.

5. Zhou, X., Lin, J., Yin, Y., Sun, X. and Tang, K., Ganodermataceae: Natural products and their related pharmacological functions. *The American J. Chinese Med.* 2007,**35**,559-574.

6. Chang, S.T. and Buswell, J.A. Mushroom nutriceuticals. *W.J.Microbiol. Biotechnol.*1998,**12**,473-476.

7. Wicks, S.M., Tong, R., Wang, C.Z., O'Connor, M., Karrison, T., Li, S., Moss, J. and Yuan, C.S., Safety and tolerability of *Ganoderma lucidum* in healthy subjects: A double-blind randomized placebo-controlled trail. *The American J. Chinese Med.* 2007, **35**,407-414.

17

Antineoplastic properties of polysaccharides from *Phellinus* species with special emphasis to *Phellinus rimosus* (Berk) Pilat

C.R. Meera[1] and K.K. Janardhanan[*2]

[1] Department of Microbiology, St Mary's College, Thrissur-680 020, Kerala, India.
[2] Department of Microbiology, Amala Cancer Research Centre, Amala Nagar, Thrissur- 680 555, Kerala, India. *E-mail: kkjanardhanan@yahoo.com

Phellinus is a large and widely distributed genus belonging to the *Hymenochaetaceae* of basidiomycetes. Many *Phellinus* species including *P. igniarius, P. hartigii, P. gilvus, P. pini,* etc. are known to have several medicinal effects such as anti-tumor and immuno-stimulating activities. About 18 species of *Phellinus* are found to occur in Kerala, most of them are wood inhabiting. *Phellinus rimosus* is a less extensively studied species of the genus *Phellinus*. Polysaccharide fraction bound to protein (PPC-*Pr*) complex isolated from the fruiting bodies of *P. rimosus* was found to possess significant biological properties such as antitumor, free radical scavenging, anti-inflammatory activities and inhibition of cell proliferation and apoptotic activity. A brief summary of antineoplastic properties of polysaccharides from *Phellinus* species is outlined in this paper.

Introduction

Mushrooms represent a major and yet largely untapped source of powerful new pharmaceutical products[1]. The number of mushroom species on earth is estimated to be 140 000, suggesting that only 10% are known. Assuming that the proportion of useful mushrooms among the undiscovered and unexamined mushrooms will be only 5%, which implies 7000 yet undiscovered species will be of possible benefit to mankind[2]. Use of fungi as medicines dates back to 3000 BC when macrofungi are used to remedy diseases by mankind especially in the traditional oriental therapies and after discovering of Penicillin (1929), fungi were regarded as rich sources of natural antibiotics and other bioactive compounds. Macrofungi such as *Ganoderma lucidum, Lentinus edodes, Fomes fomentarius, Fomitopsis officinalis* and many others have been used to remedy different diseases for hundreds of years in China, Japan, Korea and Slav regions[1]. The major immunomodulating and anti-tumor effects of mushrooms have been associated with polysaccharides, glycopeptide/protein complexes, proteoglycans, proteins, and triterpenoids[3].

Mushroom polysaccharides in the treatment of cancer

Biologically active polysaccharides are widespread among higher basidiomycetes mushrooms, and most of them have unique structures in different species. Moreover, different strains of one Basidiomycetes species can produce polysaccharides with different properties[1]. The mushroom polysaccharides have been effective against esophageal, stomach, prostate and lung cancers. Lucas et al. [4] demonstrated the anti-tumor effect of higher basidiomycetes (specifically extracts of fruiting bodies of *Boletus edulis*). Yohida et al.[5] isolated from *Lampteromyces japonicus* (kowamura) Sing, an agent active against Ehrlich carcinoma of the mouse. Gregory [6] experimented on more than 7000 cultures of higher basidiomyces for anti-tumor activity against rodent tumor systems. Positive inhibitory effects were obtained using fermentation media materials against sarcoma 180, mammary adenocarcinoma 755, and leukemia L-1210. Ikekawa et al. [7] reported that an essence obtained from the fruit body of edible mushrooms exhibited remarkable host-mediatory anti-tumor activity against grafted cancer in animals such as sarcoma 180. Daba [8] and Daba and Ezeronye, [9] reported that *Pleurotus ostreatus* mushrooms cultivated on date waste possess a potent anti-tumor activity against Ehrlich ascites carcinoma. Further, biochemical studies were carried out by the authors on the effect of mushrooms and isolated polysaccharides on the transplanted tumors in mice. The anti-tumor essence was later discovered to be a type of β-D-glucan, a polysaccharide yielding D-glucose only by acid hydrolysis[10]. The polysaccharides of mushrooms occur mostly as glucans, some of which are linked by β-(1-3), (1-6) glycosidic bonds and α-(1-3) glycosidic bonds but many are true heteroglycans. More than 50 mushroom species have yielded potential immunoceuticals that exhibit anticancer activity in vitro or in animal models. Six of these polysaccharides that have been investigated in human cancers include Lentinan, Schizophyllan, Active hexose correlated compounds (AHCC), Maitake D-fraction, polysaccharide-K (PSK) and polysaccharide-P (PSP)[9].

Biological properties of *Phellinus* species

There are approximately 220 known species of *Phellinus* mushrooms in the world. Various *Phellinus* mushrooms like *P. igniarius, P. hartigii, P. gilvus, P. pini, etc.* are known to have different medicinal effects such as anti-tumor and immuno-stimulating activities[11-15]. Polysaccharides isolated from *Phellinus gilvus* significantly inhibited benzo[*a*]pyrene-induced forestomach carcinogenesis in mice by down-regulating mutant p53 expression[16]. *Phellinus linteus* is a well-known species of the genus *Phellinus,* which attracts great attention due to its potent anti-tumor effect and other medicinal values. The anti-tumor activity of the polysaccharides from the fruiting body of this mushroom was first reported in 1968, thereafter a wide variety of further reports have been documented by many investigators[17-22]. In a comparative study of polysaccharides from Basidiomycetes, the polysaccharide of *P. linteus* was the most potent one and produced growth inhibition of 96.7% in Sarcoma 180 transplanted immunocompetent ICR mice[23]. The active polysaccharide purified from mycelia culture of *P. linteus* also stimulates humoral and cell-mediated immunity, and exhibited a wider range of immunostimulation and anti-tumor activity than other polysaccharides isolated from Basidiomycetes[24]. PL was also suggested to be used in immunotherapy of cancer because of its effective activities on tumor growth and metastasis through the immunopotentiation without toxicity[25]. Many investigators have been demonstrated that the mechanism of anti-tumor action of *Phellinus* mushroom involves multiple processes. Results of the study carried out by Kim et al. [26] suggest that PL is a biological response modifier that stimulates proliferation and expression of co-stimulatory molecules in B cells, probably by regulating protein tyrosine kinase (PTK) and protein kinase C (PKC) signaling pathways. Another study says that the tumoricidal activity of peritoneal macrophages (PM) cultured with PL against B16 melanoma cells was enhanced in a dose-dependent manner. This study concluded that PL act as an effective immunomodulator and enhances the anti-tumoral activity of PM through the up regulation of nitric oxide (NO) and Tumor necrosis factor-α (TNF- α)[27]. Extract of *P. linteus* was reported to possess antimutagenic activities and play a role in the prevention of cancer by inducing NAD(P)H:quinine oxidoreductase and glutathione S-transferase activities and increasing glutathione level[28]. Li et al.[29] reported the direct cytotoxicity of PL againt cancer cells by inhibiting the cellular proliferation of SW480 human colon cancer cells by induction of apoptosis and G_2/M phase arrest.

Medicinal properties of *Phellinus rimosus*

About 18 species of *Phellinus* are found to occur in Kerala, most of them are wood inhabiting[30]. *Phellinus rimosus* (Berk) Pilat, commonly called 'Bracket fungus' or 'Cracked cap polypore', is a polyporus macrofungus, mostly parasitic and found on jack fruit tree trunks in Kerala (Figure 1). *Phellinus rimosus* is a less extensively studied species. Basidiocarps of this mushroom have been reported to be used by some tribes in Kerala for curing mumps[31]. The significant antioxidant and anti-inflammatory activities of methanolic extract of *P. rimosus* was first reported by Ajith and Janardhanan [32]. The ethyl acetate extract of *P. rimosus* exhibited significant hepatoprotective effect against CCl_4 induced acute hepato toxicity in rats. The extract also possessed significant free radical scavenging activity in a concentration dependent manner that could exert a beneficial effect against hepatotoxicity in experimental animals[33]. The ethyl acetate extract of *P. rimosus* was also found to be effective in the amelioration of cisplatin induced nephrotoxicity in mice[34]. Ajith and Janardhanan [35] have also reported the cytotoxic and anti-tumor activities of ethyl acetate, methanol and aqueous extracts of *P. rimosus*. All the three extracts were highly effective in inhibiting growth of solid tumor induced by Dalton's Lymphoma Ascites (DLA) cell line in mice. The ethyl acetate extract was also effective in preventing the Ehrlic's Ascites Carcinoma (EAC) induced ascites tumor development in mice.

Fig. 1: *Phellinus rimosus* growing on jackfruit tree trunk

Antineoplastic properties of polysaccharides from *Phellinus rimosus*

Meera et al.[36] reported the isolation of a crude polysaccharide bound to protein from fruiting bodies of *P. rimosus* (PPC-*Pr*). The phytochemical analysis revealed that PPC-*Pr* is a D-glucan-protein complex of molecular weight 1,200,000 Da. Anti-tumor activity of PPC-*Pr* was evaluated using the EAC and DLA murine cell lines. PPC-*Pr* administration considerably increased the life span of EAC induced ascetic tumor bearing mice. PPC-*Pr* also showed significant preventive and curative effects on solid tumor induced by DLA cell line[36]. PPC-*Pr* was found to be effective in scavenging superoxide, hydroxyl, nitric oxide radicals and also the 1,1-diphenyl 2-picryl hydrazyl (DPPH) free radical. It significantly inhibited the lipid peroxidation and Ferric Reducing Anti-oxidant Power (FRAP) assay showed the reducing ability of PPC-*Pr*. The anti-inflammatory activity of the PPC-*Pr* was determined by carrageenan and dextran induced acute, and formalin and Frund's Complete Adjuvant (FCA) induced chronic inflammatory models. The PPC-*Pr* showed strong anti-inflammatory activity in both the acute and chronic inflammatory models in a dose dependent manner. The activity of PPC-*Pr* was comparable to the activity of clinically used standard drug, diclofenac[37]. The *in vivo* anti-oxidant activity of PPC-*Pr* was evaluated in FCA induced chronic inflammation in rats. Lipid peroxides formation as well as altered levels of endogenous scavengers like superoxide dismutase; glutathione peroxidase and reduced glutathione were taken as an indirect *in vivo* evidence for the participation of free radicals in the progression of chronic inflammation. The increased lipid peroxide level in plasma and the elevated activity of antioxidant enzymes such as superoxide dismutase and glutathione peroxidase were found in FCA induced chronic inflammation in rats where as, blood glutathione was decreased. The above alterations produced in FCA induced animals were brought to normal by PPC-*Pr* in a dose dependent manner[38].

Apoptosis plays an important role as a protective mechanism against carcinogenesis by eliminating damaged cells or abnormal excess cells proliferation owing to various chemical agents' induction. Emerging evidence has demonstrated that the anti-cancer activities of certain chemotherapeutic agents are involved in the induction of apoptosis, which is regarded as the preferred way to manage cancer. PPC-*Pr* was able to induce significant cytotoxicity against HCT 116 colon cancer cell line. PPC-*Pr* inhibited the proliferation of HCT 116 cells markedly at a dose of 500 and 1000μg/ml. The characteristic changes of apoptosis were observed in the morphology of PPC-*Pr* treated cells by 4', 6-Diamidino-2-phenylindole dihydrochloride hydrate and Acridine orange-ethidium bromide staining methods. Chromatin degradation into multiple internucleosomal fragments is a distinct biochemical hallmark for apoptosis and can be easily detected by Comet assay. Comet assay revealed that the apoptotic nuclei were more frequent in PPC-*Pr* treated cells than the control. Thus, by long term incubation, PPC-*Pr* induced cytotoxicity and inhibited the tumor cell proliferation in vitro by the induction of apoptosis[39].

Conclusion

Mushroom polysaccharides appear to be nontoxic in prolonged use and are claimed to benefit general health. Mushroom polysaccharides are increasingly being utilized to treat a wide variety of diseases without the need to go through phase I/II/III trials as an ordinary medicine, and they are considered as safe and useful approach for disease treatment. A wide range of biologically active polysaccharides is found among higher Basidiomycetes mushrooms and their practical application is dependent not only on their unique properties but also on biotechnological availability. Isolation of polysaccharides from *Phellinus* species is relatively simple and straight forward, and can be carried out with minimal effort. Polysaccharides isolated from this species possessed significant anti-tumor activity in different models studied. The anti-oxidant, anti-inflammatory, cytotoxic, anti-proliferative and apoptotic activities of these polysaccharides seem to contribute to their profound anti-neoplastic property. The reports suggest that the polysaccharides isolated from *Phellinus* species would be a suitable candidate for the prevention and treatment of cancer.

References

1. Wasser, S.P., Medicinal mushrooms as a source of antitumour and immunomodulating polysaccharides. *Appl. Microbiol. Biotechnol.,* 2002, **60**, 258-274.

2. Hawksworth, D.L., Mushrooms: the extent of the unexplored potential. *Int. J. Med. Mushr.,* 2001, **3**, 333-337.

3. Lucas, E.H., Montesano, R., Pepper, M.S., Hafner, M. and Sablon, E., Tumor inhibitors in Boletus edulis and other holobasidiomycetes. *Antibio. Chemother.,* 1957, **7**, 1-4.

4. Yohida, T.O., A tumor inhibitor in *Lampteromyces japonica. Pro. Soci. Experi. Bio. Med. (PSEBM).,* 1962, **3,** 676-679.

5.   Gregory, F.J., Studies on antitumor substances produced by Basidomycetes. *Mycologia.*, 1966, **58,** 80-90.

6.   Ikekawa, T., Nakanishi, M., Uehara, N., Chihara, G. and Fukuoka, F., Antitumor action of some Basidiomycetes, especially *Phellinus lintens. Gann.*, 1968, **59,** 155-157.

7.   Daba, A.S., Biochemical studies of effect of mushrooms and isolated polysaccharides on tumors transplanted in mice. *2nd Inter. Confer. Fed. Afri. Soci. Biochem. Mol. Bio., 1998,* Potshfostroom, South Africa.

8.   Daba, A.S. and Ezeronye, O.U., Anti-cancer effect of polysaccharides isolated from higher basidiomycetes mushrooms. *Afri. J. Biotech.,* 2003, **2,** 672-678.

9.   Mizuno, T., The extraction and development of antitumor active polysaccharides from medicinal mushrooms in Japan (Review). *Int. J. Med. Mush.,* 1999, **1,** 9 -29.

10. Ayer, W.A., Muir, D.J., Chakravarty, P., Phenolic and other metabolites of *Phellinus pini*, a fungus pathogenic to pine. *Phytochemistry.*, 1996, **42**, 1321-1324.

11. Jung, I.C., Kim, S.H., Kwon, Y.I., Kim, S.Y., Lee, J.S. and Park, S., Cultural condition for the mycelial growth of *Phellinus igniarius* on chemically defined medium and grains. *Kor. J. Myco.,* 1997, **25,** 133-42.

12. Rew, Y.H., Jo, W.S., Jeong, K.C., Yoon, J.T. and Choi, B.S., Cultural characteristics and fruit body formation of *Phellinus gilvus. Kor. J. Myco.,* 2000, **28,** 6-10.

13. Shibata, S., Nishikawa, Y., Mei, C.F., Fukoka, F. and Nakanishi, F., Anti-tumor studies on some extracts of Basidiomycetes. *Gann.,* 1968, **59,** 159-161.

14. Shon, Y.H. and Nam, K.S., Antimutagenicity and induction of anticarcinogenic phase II enzymes by basidiomycetes. *J. Ethnopharmacol.,* 2001, **77,** 103-109.

15. Bae, J.S., Jang, K.H., Park, S.C. and Jin, H.K., Inhibitory effects of polysaccharides isolated from *Phellinus gilvus* on benzo [*a*]pyrene-induced fore stomach carcinogenesis in mice. *World J. Gastroenterol.,* 2005, **11,** 577-579.

16. Kang, T.S., Lee, D.G. and Lee, S.Y., Isolation and mycelial submerged cultivation of *Phellinus* sp. *Kor. J. Mycol.,* 1997, **25,** 257–67.

17. Chi, J.H., Ha, T.M., Kim, Y.H. and Rho, Y.D., Studies on the main factors affecting the mycelial growth of *Phellinus linteus. Kor. J. Mycol.,* 1996, **24,** 214–22.

18. Chung, K.S., Kim, S.S., Kim, H.S., Kim, K.Y., Han, M.W. and Kim, K.H., Effect of Kp, an anti-tumor protein polysaccharide from mycellial culture of *Phellinus linetus* on the humoral immune response of tumor bearing ICR mice to sheep red blood cells. *Arch. Pharma. Res.,* 1993, **16,** 336-338.

19. Han, S.B., Lee, C.W., Jeon, Y.J., Hong, N.D., Yoo, I.D., Yang, K.H. and Kim, H. M., The inhibitory effect of polysaccharides isolated from *Phellinus linteus* on tumor growth and metastasis. *Immunopharmacol.,* 1999, **41,** 157–164.

20. Lee, J.H., Cho, S.M., Kim, H.M., Hong, N.D. and Yoo, I.D., Immunostimulating activity of polysaccharides from mycelia of *Phellinus linteus* grown under different culture conditions. *J. Microbiol. Biotechnol.,* 1996, **6,** 52–55.

21. Song, K.S., Cho, S.M., Lee, J.H., Kim, H.M., Han, S.B., Ko, K.S. and Yoo, I.D., B-lymphocyte-stimulating polysaccharide from mushroom *Phellinus linteus*. *Chem. Pharma. Bull.,* 1995, **43**, 2105–2108.

22. Kim, H.M., Han, S.B., Oh, G.T., Kim, Y.H., Hong, D.H., Hong, N.D. and Yoo, I.D., Stimulation of humoral and cell mediated immunity by polysaccharide from mushroom *Phellinus linteus*. *Int. J. Immunopharmacol.,* 1996, **18**, 295–303.

23. Han, S.B., Lee, C.W., Jeon, Y.J., Hong, N.D., Yoo, I.D., Yang, K.H. and Kim, H. M., The inhibitory effect of polysaccharides isolated from *Phellinus linteus* on tumor growth and metastasis. *Immunopharmacol.,* 1999, **41**, 157–164.

24. Kim, G.Y., Park, S.K., Lee, M.K., Lee, S.H., Oh, Y.H., Kwak, J.Y., Yoon, S., Lee, J.D. and Park, Y.M., Proteoglycan isolated from *Phellinus linteus* activated murine B lymphocytes via protein kinase C and protein tyrosine kinase. *Int. Immunopharmacol.,* 2003, **3**, 1281-1292.

25. Kim, G.Y., Choi, G.S., Lee, S.H. and Park, Y.M., Acidic polysaccharide isolated from *Phellinus linteus* enhances through the up-regulation of nitric oxide and tumor necrosis factor-α from peritoneal macrophages. *J. Ethnopharmacol.,* 2004, **95**, 69-76.

26. Kim, G.Y., Park, S.K., Lee, M.K., Lee, S.H., Oh, Y.H., Kwak, J.Y., Yoon, S., Lee, J.D. and Park, Y.M., Proteoglycan isolated from *Phellinus linteus* activated murine B lymphocytes via protein kinase C and protein tyrosine kinase. *Int. Immunopharmacol.,* 2003, **3**, 1281-1292.

27. Kim, G.Y., Choi, G.S., Lee, S.H. and Park, Y.M., Acidic polysaccharide isolated from *Phellinus linteus* enhances through the up-regulation of nitric oxide and tumor necrosis factor-α from peritoneal macrophages. *J. Ethnopharmacol.,* 2004, **95**, 69-76.

28. Shon, Y.H. and Nam, K.S., Antimutagenicity and induction of anticarcinogenic phase II enzymes by basidiomycetes. *J. Ethnopharmacol.,* 2001, **77**, 103-109.

29. Li, G., Kim, D. H., Kim, T. D., Park, B. J., Park, H. D. and Park, J. T., Protein bound polysaccharide from *Phellinus linteus* induces G2/M phase arrest and apoptosis in SW480 human colon cancer cells. *Can. Lett.,* 2004, **216**, 175-181.

30. Leelavathy, K.M. and Ganesh, P.N., Polypores of Kerala. New Delhi: Daya Publishing House; 2000.

31. Ganesh, P.N., Studies on wood inhabiting macrofungi of Kerala. *Ph.D Thesis, Calicut University, Calicut,* India., 1988.pp 45.

32. Ajith, T.A. and Janardhanan, K.K., Antioxidant and anti-inflammatory activities of methanol extract of *Phellinus rimosus. Ind. J. Exp. Biol.,* 2001, **39**, 1166-1169.

33. Ajith, T.A. and Janardhanan, K.K., Antioxidant and antihepatotoxic activities of *Phellinus rimosus* (Berk) Pilat. *J. Ethnopharmacol.,* 2002, **81**, 387-391.

34. Ajith, T.A. and Janardhanan, K.K., Amelioration of cisplatin induced nephrotoxicity in mice by ethyl acetate extract of a polypore fungus, *Phellinus rimosus. J. Exp. Clin. Can. Res.,* 2002, **21**, 487-491.

35. Ajith, T.A. and Janardhanan, K.K., Cytotoxic and antitumor activities of a polypore macrofungus, *Phellinus rimosus* (Berk) Pilat. *J. Ethnopharmacol.*, 2003, **84**, 157-162.

36. Meera, C.R. and Janardhanan, K.K., Antitumor activity of a polysaccharide-protein complex isolated from a wood-rotting polypore macrofungus *Phellinus rimosus* (Berk) Pilat. *J. Environ. Pathol. Toxicol. Oncol.*, 2012, **31**, 223-232.

37. Meera, C.R. Janardhanan, K.K., Nitha, B. and Viswakarma, R.A, Anti-inflammatory and free radical scavenging activities of polysaccharide-protein complex isolated from *Phellinus rimosus* (Berk) Pilat. *Int. J. Med. Mushr.*, 2009, **11**,365-373.

38. Meera, C.R., Smina, T.P., Nitha, B., John, M., and Janardhanan, K.K., Anti-arthritic activity of a polysaccharide-protein complex isolated from *Phellinus rimosus* (Berk) Pilat (Aphyllophoromycetideae), in Freund's complete adjuvant induced arthritic rats. *Int. J. Med. Mushr.*, 2009, **11**, 21-28.

39. Meera, C. R., Studies on the antineoplastic properties of polysaccharides isolated from a wood rotting macrofungus *Phellinus rimosus* (Berk) Pilat. Ph. D thesis, Mahatma Gandhi University, Kottayam, Kerala, India, 2009, pp 93-107.

18

South Indian medicinal mushrooms – a potential source for anticancer drugs

V. Kaviyarasan

C.A.S. in Botany, University of Madras, Guindy campus, Chennai – 600 025, Tamil Nadu, India, e-mail: manikavi53@gmail.com

Biodiversity study on mushrooms was initiated and more than seven hundred species were described from South India. But the bio-documentation of many medicinal mushrooms was initiated a decade ago. *Lentinus tuberregium, Neolentinus kauffmanii, and Agaricus heterocystis* **were studied for their medicinal properties such as antitumor, antiviral, antimicrobial and antioxidant effects. These indigenous mushrooms are effective against many cancer cell lines and induce apoptosis and results in tumor cell death. Antiangiogenic effect of** *Trametes hirsuta* **extract was well established in fertilized hen eggs. Polysaccharides from** *Tramates hirusuta,* **an indigenous isolate, were also very effective against many cancer lines. These results, clearly established their candidature for drug formulations. Two novel anticancer compounds extracted from** *L. tuberregium* **were filed for patent. Currently few more edible mushrooms are being studied for their medicinal properties.**

Introduction

Many clinically important drugs, such as aspirin, digitoxin, progesterone, cortison and morphine, have been derived directly or indirectly from higher plants. Less well-recognized but of great clinical importance are the widely used drugs from fungi such as the antibiotics, penicillin and griseofulvin, the ergot alkaloids and cyclosporine[1]. Mushrooms are the macrofungi with a distinctive fruiting body, which can be hypogeous or epigeous, large enough to be seen with the naked eye and to be picked by hand'[2]. Mushrooms constitute at least 14,000 and perhaps as many as 22,000 known species. The number of mushroom species on the earth is estimated to be 1,40,000 suggesting that only 10% are known[3].

For millennia, mushrooms have been valued by man-kind as an edible and medical resource. A number of bio-active molecules, including antitumor substances, have been identified in many mushroom species. During the last two decades there has been an increasing recognition of the role of the human immune system for maintaining good health. Mushrooms such *as Ganoderma lucidum* (Reishi), *Lentinus edodes* (Shiitake), *Inonotus obliquus* (Chaga) and many others have been collected and used for hundreds of years in Korea, China, Japan, and eastern Russia. It is notable and remarkable how reliable the facts collected by traditional eastern medicine are in the study of medicinal mushrooms [4,5]. In India, the knowledge of indigenous mushroom consumption as food and medicine prevails from time immemorial. But there is no authentic record of our own. But two authentic reports on medicinal uses were recorded on the uses of *Termitomyces* mycelial mass near Thanjavur, Tamilnadu, India and Gordon Wasson in his book on Soma drink referred in Rig Vedas stating that *Amanita muscaria* extract was the Soma drink by comparing the descriptions in the Vedas with the structural description of *Amanita*. Use of compounds from *Phellinus* sp. as preservative was recorded by Sharifi et al.[6] and their antitumor activity was studied by Meera and Janardhanan[7]. Natarajan et al.[8] has studied biodiversity of Agaric diversity of South India. Recently, many indigenous edible and medicinal mushrooms were studied by our group for their antioxidant and antitumor activities using cell lines [9-13].

Current status of research

The fruiting body and the mycelium of mushrooms contain compounds with a wide range of medicinal properties. Currently, there are lots of research have been carried out to prove the medicinal properties such as antitumor properties, antiviral, antibacterial and immunomodulatory properties of the bioactive metabolites at both national and international level. Mushrooms are rich sources of ß-glucan, proteoglucan, lectin, phenolic compounds, flavonoids, polysaccharides, triterpenoids, diatery fibre, lentinan, schizophyllan, lovastatin, pleuran, steroids, glycopeptides, terpenes, saponins, xanthones, coumarins, alkaloid, kinon, fenil propanoid, kalvasin, porisin, AHCC, maitake D-fraction, ribonucleases, eryngeolysin etc. [14-17]. Medicinal mushroom research has focused on discovery of compounds that can modulate positively or negatively the biologic response of immune cells. Those compounds, which appear to stimulate the human immune response, are being sought for the treatment of cancer, immunodeficiency disease or for generalized immunosuppression following drug treatment.

Wasser [17] reported that mushroom polysaccharides are regarded as biological response modifiers. This basically means that they cause no harm and place no additional stress on the body, but help the body to adapt to various environmental and biological stresses. Mushroom polysaccharides support some or all of the major systems of the body, including nervous, hormonal and immune systems as well as regulatory functions and produce their anti-tumour effect.

Polysaccharides are a structurally diverse class of macromolecules able to offer the highest capacity for carrying biological information due to a high potential for structural variability whereas, the nucleotides and amino acids in nucleic acids and proteins respectively interconnect in only one way. The monosaccharide units in polysaccharides can interconnect at several points to form a wide variety of branched or linear structures. This high potential for structural variability polysaccharides gives the necessary flexibility to the precise regulatory mechanisms of various cell-cell interactions in higher organisms. The polysaccharides of mushrooms occur mostly as glucans. Some of which are linked by β-(1-3), (1-6) glycosidic bonds and α-(1-3) glycosidic bonds but many are true heteroglycans.

Bioactive Polysaccharides

Historically, hot-water-soluble fractions (decoctions and essences) from medicinal mushrooms, i.e., mostly polysaccharides, were used as medicine in the Far East, where knowledge and practice of mushroom use primarily originated [5]. Ikekawa et al.[18] published one of the first scientific reports on antitumor activities of essences obtained from fruiting bodies of mushrooms belonging to the family Polyporaceae (Aphyllophoromycetideae) and a few other families, manifested as host-mediated activity against grafted cancer – such as Sarcoma 180 – in animals. Soon thereafter, the first three major drugs were developed from medicinal mushrooms. All three were polysaccharides, specifically β-glucans: krestin from cultured mycelia biomass of *Trametes versicolor* (Turkwey Tail), lentinan from fruiting bodies of *L. edodes,* and schizophyllan from the liquid cultured broth product *of Schizophyllum commune.*

Hobbs[5] reported that *L. edodes* produces two bioactive preparations, which are efficient immune modulators, mycelium extract and lentinan. These two bioactive polymers appear to act as host defense potentiator restoring and enhancing the responsiveness of host cells to lymphocytokines, hormone and other biologically active substances. The immunopotentiation has been shown to occur by stimulating the maturation, differentiation or proliferation of cells involved in host defense mechanism. Many interesting biological activities of lentinan including increase in the activation of non-specific inflammatory response such as acute phase protein production; vascular dilation and haemorrhage-inducing factor in vivo[19], activation and generation of helper and cytotoxic T cells [14].

Chihara et al.[14] reported that lentinan increase host's resistance against various kinds of cancer and has the potential to restore the immune function of affected subjects. The interaction of lentinan with many kinds of immune cells was not known until recently. Ross et al.[20] provided an insight into receptor binding in immune cells by β-glucan from fungi and further showed that β-glucan from yeast bind to iC3b- receptors (CR3, CD11b/CD18) of phagocytic and natural killer (NK) cells. When this happens, it will stimulate phagocytosis and/or cytotoxic degranulation. Lentinan has also been shown to stimulate peripheral blood lymphocytes in vitro to increase interleukin-2-mediated LAK cell (lymphokine-activated killer cell) and NK cell activity at levels achievable in vivo by administration of clinical doses of Lentinan. This observation was made using the blood of healthy donors and cancer patients. Lentinan has also been shown to inhibit suppressor T cells activity in vivo and to increase the ratio of activated T cells and cytotoxic T cells in the spleen when administered to gastric cancer patients undergoing chemotherapy.

Bioactive small molecules

Apart from Polysaccharides, mushrooms also contains valuable bioactive small molecule that have antitumor, antimicrobial and antiviral properties. The small molecule which possess anti-microbial activity includes Applanoxidic acid A, isolated from *Ganoderma annulare* (Fr.) Gilbn., shows weak antifungal activity against *Trichophyton mentagrophytes*[21]. Steroids like 5a-ergosta-7,22-dien-3b-ol or 5,8-epidioxy-5a,8a-ergosta-6,22-dien-3b-ol, isolated from *G. applanatum* (Pers.) Pat., proved to be weakly active against a number of gram-positive and gram-negative microorganisms. Oxalic acid is one agent responsible for the antimicrobial effect of *Lentinula edodes* (Berk.) Pegler against *S. aureus* and other bacteria[22]. Ethanolic mycelial extracts from *L. edodes* possess antiprotozoal activity against *Paramecium caudatum* [23]. The antimicrobial activity of *Podaxis pistillaris* (L.: Pers.) Morse, used in some parts of Yemen for the treatment of 'nappy rash' of babies and in South Africa against sun burn, is caused by epicorazins[24]. These substances belong to the group of epipolythiopiperazine-2,5-diones, an important class of biologically active fungal metabolites. Other antimicrobial compounds from the Aphyllophorales were summarized by Zjawiony[25].

In contrast to bacterial infectious diseases, viral diseases cannot be treated by common antibiotics and specific drugs are urgently needed. Antiviral effects are described not only for whole extracts of mushrooms but also for isolated compounds. They could be caused directly by inhibition of viral enzymes, synthesis of viral nucleic acids or adsorption and uptake of viruses into mammalian cells. These direct antiviral effects are exhibited especially by smaller molecules. Indirect anti-viral effects are the result of the immunostimulating activity of polysaccharides or other complex molecules.

Several triterpenes from *G. lucidum* (M. A. Curtis: Fr.) P. Karst. [i.e. ganoderiol F, ganodermanontriol, ganoderic acid B are active as antiviral agents against human immunodeficiency virus type 1 (HIV-1). The minimum concentration of ganoderiol F and gano- dermanontriol for complete inhibition of HIV-1 induced cytopathic effect in MT-4 cells is 7.8 mg/ml. Ganoderic acid B inhibits HIV-1 protease[26]. Ganodermadiol, lucidadiol and applanoxidic acid G, isolated from *G. pfeifferi*, but also known from other *Ganoderma* species, possess *in vitro* antiviral activity against influenza virus type A. Further, ganodermadiol is active against herpes simplex virus type 1, causing lip exanthema and other symptoms[27]. *In vitro* antiviral activity against influenza viruses type A and B was demonstrated for mycelial extracts of *Kuehneromyces mutabilis* (Schaeff.: Fr.) Singer & A. H. Sm.[28], extracts and two isolated phenolic compounds from *Inonotus hispidus* (Bull.: Fr.) P. Karst [29] and ergosterol peroxide, present in several mushrooms. The antiviral activity of *Collybia maculata* (Alb.&Schwein.: Fr.) P.Kumm. (vesicular stomatitis viruses in BHK cells) is caused by purine derivatives[30]. Thus many drugs are formulated not only to treat against diseases but to stimulate the immune system to resist the pathogens using biomolecules from mushrooms.

Current status of research in our laboratory

The major objective of research in our laboratory is to study the biodiversity of basidiomycetes of both Eastern Ghats (Thirumala hills, Kolli hills and Javvadi hills etc) and Western Ghats besides the plains of Tamilnadu. Besides the biodiversity study bio-documentation of these organisms is the need of the hour. Medicinal properties of few South Indian mushroom species were charcterised by isolating few bio active molecules from indigenous mushroom species. An intracellular fibrinolytic protease from *G. lucidam* isolate VK12 (Figure 1) was isolated and purified[31] which has the potential to be used as an alternative to the commercially available urokinases having many side effects on the patient for treatment of cardiovascular diseases. The enzyme was purified to homogeneity and molecular mass of 33.2 kDa. By enzyme kinetic studies, the enzyme was characterized as metalloprotease. The purified fibrinolytic protease showed anticoagulant activity with human plasma. Moreover, the purified protease protected pulmonary thromboemolism in mice to the extent of 70%. The survival rate of mice treated with purified protease were 36, 72 and 81% at doses 20, 40 and 60 µg/kg respectively, compared to 9% in the control[31].

Though some of the agaric species are growing wildly in our tropical environment, no studies were carried out to cultivate them. *Agaricus heterocystis* had formed fruit bodies in the agar medium itself [8]. Cultivation of the indigenous wild edible variety *Agaricus heterocytis strain VKJ17* (Figure 2) was later standardized by Jagadish[32]. In addition to the study on the nutritive value and their edibility and the medicinal properties, antitumor activity were also evaluated[32]. More over, the ethanolic extract of this mushroom induces apoptotic mode of cell death in HL-60 cell line. Two more terpenoid compounds were isolated namely C1-AGH and C2-AGH were shown in the Figure 3, which were active against human viruses such as HSV type1, type 2 and Influenza viruses A and B.

Lentinus tuberregium, an edible mushrooms consumed by Kaani tribes of Paechi parai forest of Western Ghats. The antitumor, antioxidant and antimicrobial diterpene compounds were isolated from the *L. tuberregium VKJM* 24 in another biodocumentation study[33]. Two compounds namely LT-1 and LT-2 were isolated, purified and their structure has been elucidated using various techniques and shown in the Fig 6. They were tested against various cell lines *viz.,* SK-OV-03 (ovarian cancer), A673 (Rhabdomyosarcoma), HCT-116 (Colorectal Carcinoma) and MCF-7 (Breast Cancer) and the viability of cells was determined by the MTT assay. Of the four cancer lines tested, both LT1 and LT2 exerted maximal growth inhibition in SK-OV-03, followed by A673. On the other hand, HCT-116 showed moderate growth inhibition.

Neolentinus kauffmanii strain *VKGJ01*was isolated during biodiversity study on Western Ghats [34] which is being consumed by the Kanni tribes of Kanyakumari forests of Westernghats as food additive and they claim many medicinal properties. Its cultivation was standardized for mass production. Both fruit body and mycelium were screened for bioactive compounds for antimicrobial and antitumor activity. A compound of steroid nature namely Betasitosterol with a molecular formula of $C_{28}H_{50}O_3$ and molecular weight 414.71 g/mol[-1] was isolated from *N. kauffmanii* and studied for anticancer activity. The study clearly showed that the compound exerted significant inhibitory effect on the HepG2 lung cancer cell lines. Both crude extract of mycelium and fruit body exhibited similar activities[35]. A Polysaccharide of glucan nature with antiangiogenic property was isolated from *Pleurotus eryngii* [36].

Besides these agaric members a polypore namely *Trametes hirusuta* was isolated from suburb of Chennai and its cultivation was standardized for the fruit body production. Though extensive studies on *Trametes versicolor* were carried out and the extracellular polysaccaride was shown to have antitumor activity and marketed globally as Krestin an anticancer drug[17]. Since, no such studies were carried out on indigenous *T. hirusuta* a study was carried out for biodocumetation of this mushroom[36]. An extracellular β- glucan was isolated from the *T. hirsuta strain VKESR* culture filtrate with moderate *in vitro* antioxidant and immunomodulatory potentials and showed good antiproliferative activity in colon, liver and leukemic cells lines. Further, it induced apoptosis through intrinsic mitochondrial mediated pathway in the cell lines. Moreover, it possesses good *ex vivo* antiangiogenic and has good anticancer potentials against diethyl nitrosamine-induced hepatocellular carcinoma in rats[36].

Commercial status of biomolecules

At present, between 80 to 85% of all medicinal mushroom products are derived from the fruiting bodies, which have been either commercially farmed or collected from the wild mushrooms. Only 15% of all products are based on extracts from mycelia. Examples are PSK and PSP from *T. versicolor* and Tremellastin from *Tremella mesenterica*. (Retzius): Fr. A small percentage of mushroom products are obtained from culture filtrates, e.g. Schizophyllan from *S. commune* and protein-bound polysaccharide complex from *Macrocybe lobayensis* (R. Heim) Pegler & Lodge [syn. *Tricholoma lobayense* R. Heim]. After production, suitable galenic formulations like capsules, tablets or teas have to be developed, dependent on the material. Mixtures of several mushrooms or of mushroom and substrate become more and more common.

Lentinan from *L. edodes* fruit-bodies, Schizophyllan from *Schizophyllum commune* mycelial broth, PSK and PSP, from mycelial cultures of *Trametes versicolor* and Grifron-D from fruit-bodies of *Grifola frondosa* were clinically tested commercial anticancer and immunomodulating drugs. All have been gone through Phase I, II and III clinical trials mainly in Japan and China but now in US. However, in many cases the standards of these trials may not meet current Western regulatory requirements. In many cases, there have been significant improvements in quality of life and survival. Increasingly, several of these compounds are now used extensively in Japan, Korea and China, as adjuncts to standard radio and chemotherapy. While most of these clinical studies have used extracts from individual medicinal mushrooms, some recent studies from Japan have shown that mixtures of extracts from several known medicinal mushrooms, when taken as a supplement, have shown beneficial effects on the quality of life for some advanced cancer patients.

Future focus in the relevant field

The above studies clearly show that mushrooms, similar to plants, have a great potential for the production of useful bioactive metabolites and they are a

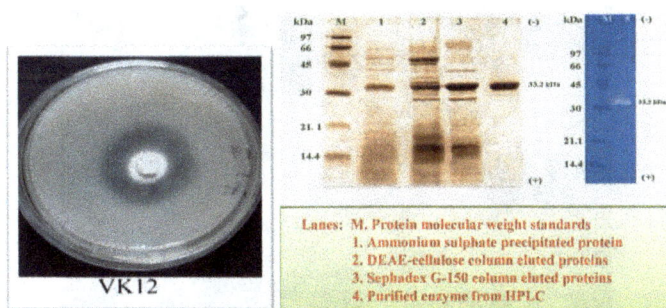

Fig. 1: Extracellular fibrinolytic protease from *Ganoderma lucidum* strain VK-12 1a) Mycelium showing fibrinolytic activity on agar plate 1b) Purification of fibrinolytic protease from *Ganoderma lucidum* strain VK-12.

Fig. 2: Fruiting body of Indigenous agaric *Agaricus heterocystis* Heinem and Gooss – *VKJ 17*

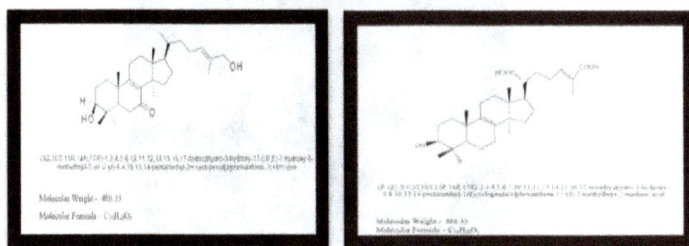

Fig. 3: Two antiviral compounds active against influenza virus A and B and HSV Type I and Type II viruses obtained from *Agaricus heterocystis* strain VKJ-17 3a) C1-AGH 3b) C2-AGH

Fig. 4: Fruiting body of *Lentinus tuberregium* (Fr.) Fr strain VKJM 24

(5a)

(5b)

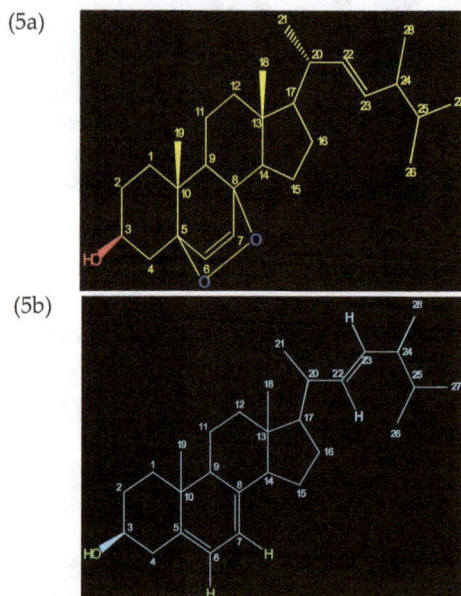

Fig. 5: Structure of purified compounds from *Lentinus tuberregium* (5a) LT-1 and (5b) LT-2. Both showing antimicrobial and antiproliferative activity

Fig. 6: Fruiting body of *Neolentinus kauffmanii* Smith A.H. strain VKGJ01

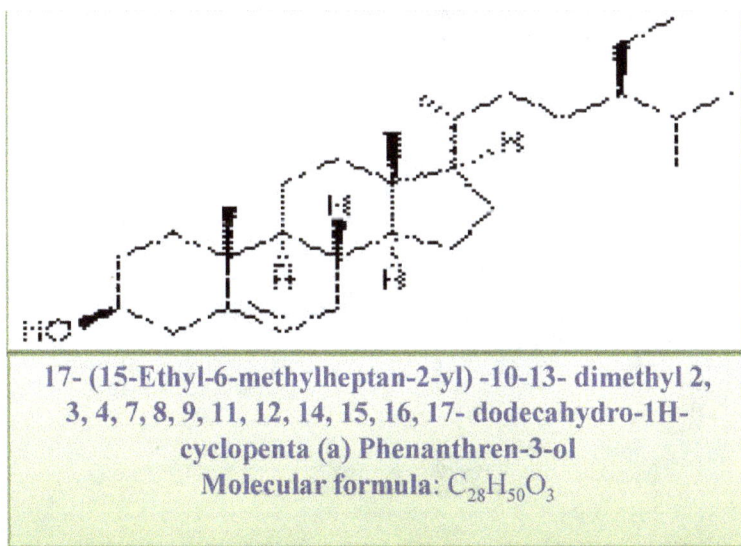

17- (15-Ethyl-6-methylheptan-2-yl) -10-13- dimethyl 2,
3, 4, 7, 8, 9, 11, 12, 14, 15, 16, 17- dodecahydro-1H-
cyclopenta (a) Phenanthren-3-ol
Molecular formula: $C_{28}H_{50}O_3$

Fig. 7: Structure of purified anticancer compound NK-1 active against liver cancer cell lines obtained from *Neolentinus kauffmanii* strain VKGJ01

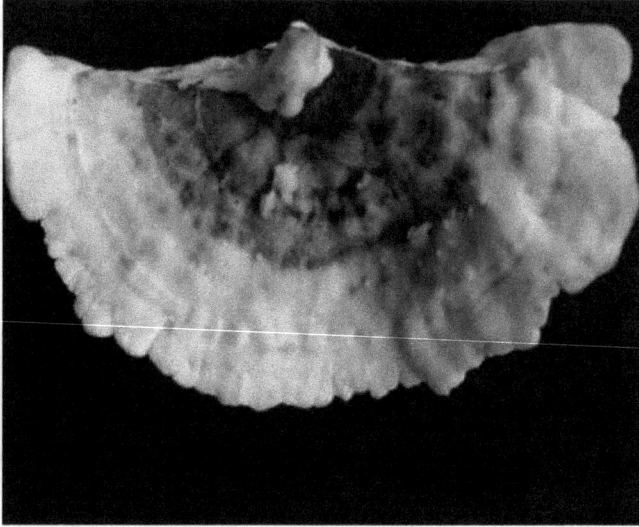

Fig. 8: Fruiting body of *Trametes hirsuta* (Wulf.) Pil. strain VKESR

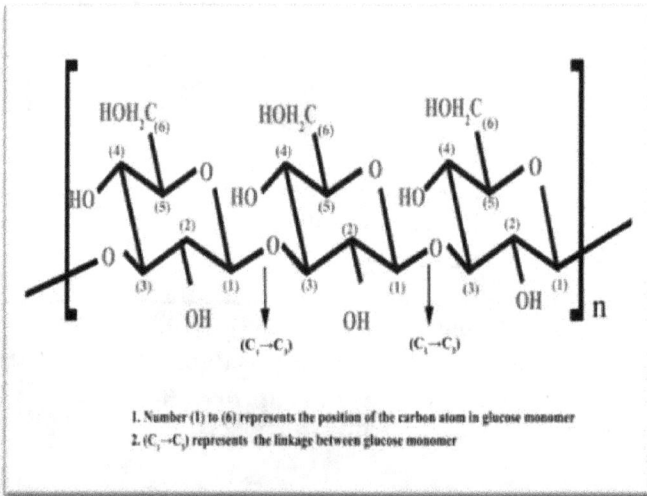

1. Number (1) to (6) represents the position of the carbon atom in glucose monomer
2. ($C_1 \rightarrow C_3$) represents the linkage between glucose monomer

Fig. 9: Structure of antitumor and immunomodulatory extracellular β – Glucan
obtained from *Trametes hirsuta* (Wulf.) Pil. strain VKESR

prolific resource for drugs. The responsible bioactive compounds belong to several chemical groups, very often they are polysaccharides or triterpenes. One species can possess a more variety of bioactive compounds, and therefore of having higher pharmacological effects. The best example is *G. lucidum*, which not only contains >120 different triterpenes but also polysaccharides, proteins and other bioactive compounds. However, one main pre-requisition for a use as drug, nutraceutical or other purpose is the continuous production of mushrooms (fruiting bodies or mycelium) in high amounts and in a standardized quality. Mycelia products are the 'wave of the future' because they ensure standardized quality and year around production. A further necessity is the establishment of suitable quality parameters and of analytical methods to control these parameters. Nevertheless, the legal regulations for authorization as drug or as dietary supplements or as food should get more attention. Control of possible side effects (i.e. allergies) during broad use is necessary. Finally, also the nutritional value of mushrooms should be taken into account.

Currently, biodocumentation of few more indigenous mushrooms from Indian forest ecosystem are being carried out in our group. Focusing on standardization of mass production of fruit bodies and mycelium for more compound production to develop novel anticancer drugs from the indigenous mushrooms in collaboration with various organizations, many more miles to go before we get a fruitful drug with following vision and concepts. More studies are need to demonstrate anti-viral, anti-tumor and hypocholesterolemic properties with high-quality long term double-blinded placebo-controlled studies with large trial populations are definitely needed for safety and efficacy of medicinal mushrooms with statistical power.

Acknowledgement

I thank the former Directors of C.A.S in Botany Prof. N. Anand, and the present Director Prof. R. Rengasamy for the help and guidance.

References

1. Ajíth, T.A. and Janardhanan, K.K., Indian medicinal mushrooms as a source of antioxidant and antitumor agents', *J. Clin. Biochem. Nutr.* 2007,**40,**157-162.

2. Chang, S.T. and Miles, P.G. *Edible mushrooms and their cultivation*, 1989,CRC Press.

3. Hawksworth, D.L. The magnitude of fungal diversity: the 1.5 million species estimate revisited, *Mycological Res.* 2001;**105,**1422-1432.

4. Wasser, S.P. and Weis, A.L. Therapeutic effects of substances occurring in higher Basidiomycetes mushrooms: a modern perspective. *Crit Rev Immunol,* 1999,**19.**65-96.

5. Hobbs, C., Medicinal value of *Lentinus edodes* (Berk.) Sing. (Agaricomycetideae). A literature review', *Int. J. Med. Mushr.* 2000,**2,**287-302.

6. Sharifi, A., Bhosle, S.R. and Vaidya, J.G., Evaluation of crude sesquiterpenoid extract of Phellinus fastuosus as a natural preservative, *Hindustan Antibiot Bull,* 2005,**47-48,** 20-23.

7.   Meera, C.R. and Janardhanan, K.K., Antitumor Activity of a Polysaccharide-Protein Complex Isolated from a Wood-Rotting Polypore Macro Fungus *Phellinus rimosus* (Berk) Pilat, *J Environ Pathol Toxicol Oncol.* 2012,**31,** 223-232

8.   Natarajan, K., Narayanan, K. and Kumaresan. V., Basidiocarp production in *Agaricus heterocystis* Heinem and Gooss in nutrient agar media. *Curr. Sci.* 2002,**82,** 9-10.

9.   Jagadish, K.L., Venkata krishnan,V., Shenbhagaraman, R and Kaviyarasan,V., Comparitive study on the antioxidant, anticancer and antimicrobial property of *Agaricus bisporus* (J. E. Lange) Imbach before and after boiling', *African J. Biotechnol.* 2009,**8,**654-661.

10. Kumaran, S., Palani, P., Nishanthi, R. and Kaviyarasan, V., Studies on screening, isolation and purification of a fibrinolytic protease from an isolate (VK12) of *Ganoderma lucidum* and evaluation of its antithrombotic activity', *Med Mycol J.* 2011, **52,**153-162.

11. Manjunathan, J., Kumar, M. and Kaviyarasan, V., Taxonomic Studies on *Lentinus tuberregium* (GQ292711) Tamil Nadu, India. Research J. *Pharm. Tech.* 2011,4.

12. Shenbhagaraman, R., Jagadish, L.K., Premalatha, K. and Kaviyarasan, V., Optimization of extracellular glucan production from *Pleurotus eryngii* and its impact on angiogenesis. *Int J Biol Macromol.* 2012,**50,**957-964.

13. Johnsy, G. and Kaviyarasan, V., Nutritive value of edible wild mushrooms collected from the Western Ghats of Kanyakumari district. *W.J.Dairy and Food Sci.* 2012, **11,**

14. Chihara, G., Immunopharmacology of Lentinan, a polysaccharide isolated from *Lentinus edodes*: its applications as a host defence potentiator. *Int. J.Oriental Med.* 1992,**17,** 57-77.

15. Wang, H., Gao, J. and Ng, T.B. A new lectin with highly potent antihepatoma and antisarcoma activities from the oyster mushroom *Pleurotus ostreatus.Biochem. Biophys. Res. Commun.* 2000,**275,** 810-816.

16. Bobek, P., Nosálová, V. and Cerná, S., Effect of pleuran (beta-glucan from *Pleurotus ostreatus*) in diet or drinking fluid on colitis in rats. *Nahrung.* 2001,**45,**360-363.

17. Wasser, S.P., Medicinal mushrooms as a source of antitumor and immunomodulating polysaccharides', *Appl. Microbiol.Biotechnol.* 2002,**60,**258-274.

18. Ikekawa, T., Uehara, N., Maeda, Y., Nakanishi, M. and Fukuoka, F., Antitumor Activity of Aqueous Extracts of Edible Mushrooms. *Cancer Res.*1969,**29,** 734-735.

19. Maeda, Y.Y., Sakaizumi, M., Moriwaki, K. and Yonekawa, H., Genetic control of the expression of two biological activities of an antitumor polysaccharide, lentinan. *Int. J. Immunopharmacol.*1991,**13,** 977-986.

20. Ross, G.D., Vetvicka, V., Yan, J., Xia, Y. and Vetvicková, J., 'Therapeutic intervention with complement and beta-glucan in cancer. *Immunopharmacol.* 1999,**42,**61-74.

21. Smania, E.F.A., Delle Monache, F., Smania Jr, A., Yunes, R.A. and Cuneo, R.S. Antifungal activity of sterols and triterpenes isolated from *Ganoderma annulare*. *Fitoter.* 2003,**74**, 375-377.

22. Bender, S., Dumitrache-Anghel, C.N., Backhaus, J., Christie, G., Cross, R.F., Lonergan, G.T. and Baker, W.L., A Case for Caution in Assessing the Antibiotic Activity of Extracts of Culinary-Medicinal Shiitake Mushroom *Lentinus edodes* (Berk.) Singer. (Agaricomycetideae). 2003,**5**,6.

23. Badalyan, S., Antiprotozoal Activity and Mitogenic Effect of Mycelium of Culinary-Medicinal Shiitake Mushroom *Lentinus edodes* (Berk.) Singer (Agaricomycetideae)', 2004,**6**, 8.

24. Lindequist U., Schizophyllum., In: Schneider G, Hänsel R, Blaschek W, editors. HAGERs Handbuch der Pharmazeutischen Praxis. Berlin, Heidelberg, Springer-Verlag, New York. 1998,528–34.

25. Zjawiony, J.K., Biologically active compounds from aphyllophorales (polypore) fungi. *J Natural Products*, 2004,**67**, 300-310.

26. El-Mekkawy, S., Meselhy, M.R., Nakamura, N., Tezuka, Y., Hattori, M., Kakiuchi, N., Shimotohno, K., Kawahata, T. and Otake, T., Anti-HIV-1 and anti-HIV-1-protease substances from *Ganoderma lucidum*. *Phytochem.* 1998,**49**, 1651-1657.

27. Mothana, R.A.A., Awadh Ali, N.A., Jansen, R., Wegner, U., Mentel, R. and Lindequist, U., Antiviral lanostanoid triterpenes from the fungus *Ganoderma pfeifferi*. *Fitoter.* 2003,**74**,177-180.

28. Mentel, R., Meinsen, D., Pilgrim, H., Herrmann, B. and Lindequist, U., In vitro antiviral effect of extracts of Kuehneromyces mutabilis on influenza virus. *Pharmazie*, 1994,**49**, 859-860.

29. Awadh Ali, N.A., Mothana, R.A., Lesnau, A., Pilgrim, H. and Lindequist, U., Antiviral activity of Inonotus hispidus. *Fitoter.* 2003, **74**,483-485.

30. Leonhardt, K., Anke, T., Hillen-Maske, E. and Steglich, W., 6-Methylpurine, 6-methyl-9-beta-D-ribofuranosylpurine, and 6-hydroxymethyl-9-beta-D-ribofuranosylpurine as antiviral metabolites of Collybia maculata (Basidiomycetes). *Zeitschrift fur Naturforschung. C, J Biosci.* 1987,**42**, 420-424.

31. Kumaran. S., Studies on Production, Purification and Characterisation of Fibrinolytic and Antithrombotic protease from a South Indian Isolate of *Ganoderma lucidum* (FR.) Karst. 2008, *Ph.D. Thesis* submitted to University of Madras, Chennai, India.

32. Jagadish, L., Cultivation and medicinal properties of indigenous Agaric. *Agaricus heterocystis* Heinem and Gooss – *VKJ 17*. 2010, *Ph.D. Thesis* submitted to University of Madras, Chennai, India.

33. Manjunathan, J., Isolation, Purification, Characterization and Evaluation of Antimicrobial and Anticancer Compounds from Indigenous Isolate of *Lentinus tuberregium* (Fr.) Fr. 2011, *Ph.D. Thesis* submitted to University of Madras, Chennai, India.

34. Sargunam, S.D., Johnsy, G., Samuel A.S. and Kaviyarasan, V., Mushrooms in the food culture of the Kaani tribe of Kanyakumari district. Indian Knowledge of traditional Medicine. 2012, **11**, *150*-153.

35. Johnsy G., Studies on nutritional and medicinal value of an indigenous isolate of *Neolentinus kauffmanii* Smith A.H. (VKGJ-01). 2012, *Ph.D. Thesis* submitted to University of Madras, Chennai, India.

36. Shenbhagaraman. R., Therapeutic potential of extracellular β- glucan from *Trametes hirsuta* (Wulf.)Pil. strain VKESR against malignant neoplasm.2012, *Ph.D. Thesis* submitted to University of Madras, Chennai, India.

19

Medicinal mushrooms as a source of immunomodulatory compounds

N.M. Krishnakumar *, P.G. Latha, S.R. Suja and S. Rajasekharan

Ethnomedicine and Ethnopharmacology Division, Jawaharlal Nehru Tropical Botanic Garden and Research Institute (JNTBGRI), Palode, Thiruvananthapuram-695 562. E-mail: krishnakumarnmohandas@gmail.com

Numerous polysaccharides and polysaccharide-protein complexes have been isolated from mushrooms with immunomodulatory and anticancer effects. The β-D-glucans like Lentinan from *Lentinus edodes*, Ganoderan from *Ganoderma lucidum*, Schizophyllan from *Schizophyllum commune*, Grifolan from *Grifola frondosa* and protein bound polysaccharides PSK (Krestin) and PSP form *Trametes versicolor* have been proved to be potent immunomodulatory compounds. The primary target of the immunomodulatory compound is believed to be the macrophages, which displaying not only increased phagocytosis and intracellular killing of pathogens but also produce cytokines like tumour necrosis factor-α (TNF-α) and interleukins IL-1, Il-6, IL 12 etc. which in turn activate T cells or NK cells. The polysaccharides and polysaccharide-protein complexes are suggested to enhance cell-mediated immune responses *in vivo* and *in vitro* and act as biological response modifiers.

Introduction

Mushrooms are an important natural source of food, medicine and nutraceuticals which are highly nutritive. They comprise a vast and yet largely untapped source of powerful pharmaceutical products. The active constituents found in mushrooms are polysaccharides, oligosaccharides, peptides etc. These have been found to boost the immune system. They are highly nutritive, low-calorie food with good quality proteins, vitamins and minerals. Mushrooms have been found effective against cancer, stress, insomnia, asthma, allergies and diabetes.The active constituents found in mushrooms are polysaccharides, dietary fibres, oligosaccharides, triterpenoids, peptides and proteins, alcohols and phenols, and mineral elements such as zinc, copper, iodine, selenium and iron, vitamins, amino acids etc. Many basidiomycetes mushrooms contain biologically active polysaccharides in fruit bodies, cultured mycelium, culture broth etc.

Immunomodulators are substances, which stimulate, suppress or modulate any one of the component of immune system. Medicinal mushroom research has focused discovering compounds that can modulate positively or negatively the biologic response of immune cells. Those compounds which appear to stimulate the human immune response are being sought for the treatment of cancer, immunodeficiency diseases or for generalized immunosuppression following drug treatment; for combinational therapy with antibiotics; and as adjuncts for vaccines[1].

A large number of chemical compounds which have been identified as specific agents for killing cancer cells are also toxic to normal cells. Many of the potential anticancer drugs have been considerable side effects. Hence the discovery and identification of new safe drugs which are active against tumours become an important goal of research in biomedical sciences. The enhancement or potentiation of host defense mechanism energies as a possible means inhibiting tumour growth without harming the host. Starting from this point of view, extensive studies have been made on polysaccharides extracted from microorganism and plant sources.

Numerous polysaccharides and polysaccharide protein complexes have been isolated from mushrooms and used as a source of therapeutic agents. The most promising bio-pharmacological activities of these biopolymers are their immunomodulation and anticancer effects[2].

Biopolymers having immunomodulatory and anticancer effects

The biopolymers mainly present as glucans with different types of glycosidic linkages such as (1-3), (1-6)-β-glucans and (1-3)-α-glucans, and as true heteroglycans, while others mostly binds to protein residues as polysaccharide-protein complexes. Three immunomodulatory and antitumour mushroom polysaccharides namely lentinan, schizophyllan and protein-bound polysaccharide (PSK, Krestin) isolated respectively from *Lentinus edodes*, *Schizophyllan commune* and *Coriolus versicolor*, have become large market item. Lentinan and schizophyllan are pure β-glucans whereas PSK is a protein-bound β-glucan. A polysaccharide peptide (PSP), isolated from a strain of *C. versicolor* in China, has also been widely used as an anticancer and immunomodulatory agent. These polysaccharides and polysaccharide complexes are suggested to enhance cell-mediated immune responses in vivo and in vitro and act as biological response modifiers. Potentiation of the host defense system may result in the activation of many kinds of immune cells that are vitally important for the maintenance of homeostasis.

Polysaccharides or polysaccharide complexes are considered as multi-cytokine inducers that are able to induce gene expression of various immunomodulatory cytokines and cytokine receptors. Some interesting studies focus on investigation of the relationship between their structure and antitumour activity, elucidation of their antitumour mechanism at the molecular level and improvement of their various biological activities by chemical modifications[3].

Mushrooms with immunomodulatory and anticancer activities

Lentinus edodes (Berk.) Singer

It is an edible mushroom native to East Asia, which is cultivated and consumed in many Asian countries. It comes under the family Marasmiaceae. Lentinan, a cell wall constituent extracted from the fruiting body with a molecular weight of about 500 Kilo Daltons (KDa). Primary structure is a (1→3)-β-glucan, consisting of five (1→3)-β-glucose residues in a linear linkage and two (1→6)-β-glucopyranoside branches in side chain which result in a right handed triple helical structure[4]. The configuration of the glucose molecules in a helix structure is thought to be important for the biological activity. Lentinan does not attack cancer cells directly, but produces its antitumour effect by activating different immune responses in the host. Evidence suggests that this immune-potentiation occurs by stimulating the maturation, differentiation or proliferation of cells involved in host defence mechanisms. The cell mediated immune response against the target cells initiated by macrophage-lymphocyte interaction and cytotoxicity induced by antibodies to target cells are believed to contribute to the elimination of target tumour cells. The induction in the amounts of IL-1, IL-3 etc. by lentinan results in maturation, differentiation and proliferation of immune-competent cells for host defense mechanism[5]. Thus, Lentinan has been shown to increase host resistance against various kinds of cancer and has the potential to restore the immune function of affected individuals.

Schizophyllum commune Fr.:Fr.

This is a small, whitish mushroom with no stalk which grows on dead trees throughout the year comes under the family Schizophyllaceae. Schizophyllan, a (1→3) β- glucan, isolated from culture filtrates of *S. commune*. It has molecular weight of about 450 K Da, having a β-glucopyranosyl group linked 1→6 to every third or fourth residue of the main chain. The polysaccharide schizophyllan exhibits antitumor activity against both the solid and ascite forms of Sarcoma 180, as well as against the solid form only of sarcoma 37, Erlich sarcoma, Yoshida sarcoma and Lewis lung carcinoma[6]. Schizophyllan has also increased cellular immunity by restoring suppressed killer-cell activity to normal levels in mice with tumors[7]. Best results against radiation damage were found when schizophyllan was administered shortly after or at the same time as radiation, and Schizophyllan restored mitosis of bone marrow cells previously suppressed by anticancer drugs[8]. Human clinical studies proved the beneficial activity of treatment with schizophyllan for patients with recurrent and inoperable gastric cancer, stage 2 cervical cancer, and advanced cervical carcinoma. Schizophyllan restores and enhances cellular immunity in the tumour-bearing host by functioning as a T-cell adjuvant and macrophage activator.

Ganoderma lucidum (Curtis) P. Karst

It has been used extensively as "mushrooms of immortality" in China and other Asian countries. *Ganoderma lucidum* comes under the family Ganodermataceae. Ganoderan is isolated from the fruiting body. Ganoderan has been shown to increase the expression of MHC class II molecules on these antigen-presenting macrophages[9.] There is also evidence to suggest that extracts from *G. lucidum* can influence humoral or B cell immunity. An alkali extract from *G. lucidum* activated both the classical and alternative pathways of the complement system. This extract also activated the reticulo-endothelial system and increased haemolytic plaque forming cells in the spleen of mice[10]. Various substances from *G. lucidum* (e.g., polysaccharides, triterpenoids, and proteins) have been shown to have marked immuno-modulating effects such as augmenting the activity of effector T cells, NK cells and macrophages.

Trametes versicolor (L.) Lloyd

A protein bound polysaccharide- PSP, isolated from cultured mycelium of *Trametes versicolor*. It is a $(1{\to}4)$-β-glucan with $(1{\to}6)$-β-glucopyranosidic side chains for every fourth glucose unit. PSK (Krestin) is a β-glucan proten complex containing 25-35 % protein residues present in *T. versicolor*, which consists of predominantly acidic aminoacids such as aspartic and glutamic acids and neutral amino acids such as lysine and arginine are present only in small amounts. The major constituent monosaccharide is glucose with small amounts of other sugar residues such as mannose, fucose, xylose and galactose. PSK has a structure with branches at 3- and 6- positions in a propotion of one per every several residual groups of $1{\to}4$ bonds[11].

Compound structure and bioactivity

Some interesting studies focus on investigation of the relationship between the structure of active compounds isolated from mushrooms and bioactivity. It is obvious that structural features such as β-$(1{\to}3)$ linkages in the main chain of the glucan and additional β-$(1{\to}6)$ branch points are needed for antitumor activity. High molecular weight glucans appear to be more effective than that of low molecular weight[12]. A triple helical conformation of β-$(1{\to}3)$-glucan is known to be important for the immune-stimulating activity. When lentinan was denatured with dimethyl sulphoxide or urea, tertiary structure was lost while primary structure was not affected, but tumour inhibition properties were lowered with progressive denaturation. Chemical modification is often carried out to improve the antitumour activity of polysaccharides and their clinical qualities.

Conclusion

Isolation and purification of polysaccharides from mushroom material is relatively simple and straight forward, and can be carried out with minimal effort. Polysaccharides from mushrooms do not attack cancer cells directly, but produce their antitumor effects by activating different immune responses in the host through a thymus-dependent immune mechanism. The antitumour activity of polysaccharides requires an intact T-cell component; their activity is mediated through a thymus-dependent immune mechanism. Mushroom polysaccharides are known to Stimulate NK cells, T-cells, B-cells and macrophage-dependent immune system responses. The immunomodulatory action of mushroom polysccharide is especially valuable as a means of prophylaxis, a mild and non-invasive form of treatment, prevention of metastatic tumors and as a co-treatment with chemotherapy.

Mushrooms have tremendous medicinal, food, drug and mineral values. They are valuable asset for the welfare of humans which substantiate the saying of Hippocrates, the father of medicine that "Let food be your medicine and medicine be your food".

References

1. Jong, S. C. and Birmingham, J. M., The medicinal value of the mushroom *Grifola*. *World J Microbiol Biotechnol.*, 1990, **6**, 101–127.

2. Ooi, V. E. C. and Liu. F., A review of pharmacological activities of mushroom polysaccharides. *Int J Med Mushr*, 1999, **1**, 195–206.

3. Ooi, V. E. C. and Liu. F., Immunomodulation and anticancer activity of polysaccharide-protein complexes. *Curr Med Chem.*, 2000, **7**, 715-729.

4. Chihara, G., The antitumor polysaccharide Lentinan: an overview. In *Manipulation of host defense mechanisms* (eds. Aoki T et al), Excerpta Med, Int Congr Ser, Elsevier, Amsterdam, 1981, pp. 576.

5. Chihara, G., Immunopharmacology of Lentinan, a polysaccharide isolated from *Lentinus edodes*: its applications as a host defence potentiator. *Int J Orient Med*, 1992, **17**, 57-77.

6. Hobbs, C., *Medicinal mushrooms: an exploration of tradition, healing and culture*. Botanica Press, Santa Cruz, Calif. 1995.

7. Borchers, A. T., Stern, J. S., Hackman, R. M., Keen, C. L. and Gershwin, E. M., Mushrooms, tumors, and immunity. *Soc Exp Biol Med.*, 1999, **221**, 281–293.

8. Zhu, D., Recent advances on the active components in Chinese medicines. *Abstr Chin Med*, 1987, **1**, 251–286.

9. Oh, J. Y., Cho, K. J. and Chung, S. H., Activation of macrophages by GLB, a protein polysaccharide of the growing tips of *Ganoderma lucidum*. *Yakhak Hoeji*, 1998, **42**, 302-306.

10. Lee, J. W., Chung, C. H., Jeong, H. and Lee, K. H., Effects of alkali extract of *Ganoderma lucidum* IY007 on complement andres. *Kor J Mycol*, 1990, **18**, 137-144.

11. Tsukagosihi, S., Hashimoto, Y., Fujii, G., Kobayashi, H., Nomoto, K. and Orita, K., Krestin (PSK). *Cancer Treat Rev*, 1984, **11**, 131-155.

12. Mizuno, T., Medicinal effects and utilisation of *Cordyceps* (Fr. Link (Ascomycetes) and *Isaria* Fd. (Mitosporic Fungi) Chinese caterpillar fungi "Tochukaso" (Review). *Int J Med Mushrooms*, 1996, **1**, 251-261.